JUN 27 2009

A CHRONOLOGY
OF WEATHER

DANGEROUS WEATHER

A CHRONOLOGY
OF WEATHER

Michael Allaby

Facts On File, Inc.

W

A CHRONOLOGY OF WEATHER

Facts On File, Inc.
11 Penn Plaza
New York NY 10001

Library of Congress Cataloging-in-Publication Data

Allaby, Michael.
    A chronology of weather / Michael Allaby.
      p. cm.—(Dangerous weather)
    Includes bibliographical references and index.
    ISBN 0-8160-3521-0 (alk. paper)
    1. Weather—History. 2. Weather—Chronology. 3. Natural disasters—Chronology. I. Title. II. Series:
    Allaby, Michael.
    Dangerous weather.
    QC981.A55 1998
    363.34'92'09—dc21   97-29898

Text design and illustrations by Richard Garratt
Layout by Robert Yaffe
Cover design by Matt Galemmo

Printed in the United States of America

RRD FOF 10 9 8 7 6 5 4

This book is printed on acid-free paper.

Contents

Exploring the weather

All of us are affected by the weather, but most of the time it is fairly ordinary. It may be fine or wet, warm or dry, but it is not usually exceptional. Forecasting the weather is difficult in detail, but in general weather is quite predictable. People will look at you oddly if you try to warn them that the winter will be cold, or that summer is the best time for swimming outdoors because it is warmer then.

Now and again, however, the weather does something extraordinary, and then it can be dangerous. This series is about ways the weather can behave badly. The other books in the series describe hurricanes, tornadoes, blizzards, drought, and floods, one book dealing with each. These are examples of extreme weather; all of them cause hardship, destruction, and death. In the past they have done so on a vast scale. Today we are able to limit the harm they do by warning people in advance and providing efficient emergency services that can respond quickly. Even so, extreme weather remains dangerous because its physical power is huge.

Each book in the series fully describes the harm hurricanes, tornadoes, blizzards, droughts, and floods have caused in the past and are still capable of causing. The books also tell you what is being done to protect lives and property, and what you and your family should do to prepare and how to react should one of these monsters strike in your area. Dangerous weather can injure and kill people, but it is possible to survive unharmed. Survival depends on knowing what to expect, making a sensible plan long before the weather emergency occurs, and then, when it does, carrying out your plan without panic. There will be experts to advise and help you: The National Weather Service will keep you informed, and the emergency services will issue instructions.

These extremes may be dangerous, but they are produced by the same natural forces that produce our ordinary weather. They are, after all, wind, snow, dry weather, and rain. To understand how they occur, therefore, you need to know something about those forces. You need to understand how air moves over the surface of the Earth, what makes the wind blow, how clouds form, and why it rains or snows. As you read about dangerous weather, you will also learn about ordinary weather.

This book is different from the others in this series. It starts by explaining how the climates of the world have changed over the centuries. Because the weather we experience is much the same from one year to the next, apart from rare extremes, we tend to think it has always been much as it is now. If we imagine, say, a Roman soldier or a medieval knight, we think of him living in a climate just like that of today. This is wrong. The weather then was different. It has changed many times, and it is changing now. After discussing climate change throughout history, the book outlines the way our understanding of the weather and ability to forecast it have developed over the centuries.

Scientists try to explain the phenomena they observe and then devise experiments to test their explanations. Nowadays, these experiments are usually complicated and call for very expensive equipment, but the principles behind them are simple; many features of the weather can be easily reproduced on a small scale. The 30 experiments described in the book are easy to perform and require equipment you can find around the house or improvise from ordinary materials. You may find they help you understand ideas that can be difficult to grasp.

The largest part of the book consists of two chronological accounts. One lists some of the major weather disasters by the years in which they occurred. These become more frequent in modern times, but that is because we have more detailed records of them. There is no reason to suppose weather disasters happen more often now than ever before. Look at how many hurricanes, tornadoes, blizzards, droughts, and floods there were between 1990 and 1996, at the appalling harm they have done and the personal tragedies they have caused, and remember that these years are not special. This is how the weather has behaved in any similar period throughout history, as far back as you care to delve, but we know nothing about those more distant events. Remembering this, try to imagine how terrifying dangerous weather must have been to people who had no weather service to warn them or clear ideas about what caused these disasters. Notice, too, that no part of the world escapes. Tropical cyclones form only in the tropics, but they can sometimes survive long enough to cause serious damage and loss of life as far away as Canada and Europe, and those who live in middle latitudes suffer winds of hurricane force and blizzards. The United States has more tornadoes than any country, but every country experiences some.

The book's other chronological account, listing important developments in the understanding of weather and improvements in forecasting, is not limited in the same way as the first chronology by lack of recorded information. These advances are also listed by the years in which they occurred, and our knowledge of them is much more complete, at least over the last few centuries, because they were recorded.

Together, these six books will introduce you to the adventure of exploring weather. They may even persuade you to carry your studies further and one day join the professional meteorologists, with their balloons, instruments, satellites, and powerful computers.

History of weather

Weather affects us all. We may want to know whether we can safely invite friends to a picnic next weekend, or whether the ground will be dry enough for the game we hope to see or play in a few days' time. It all depends on the weather.

Different people are affected in different ways, of course, and for some, weather forecasting is a more serious matter than being able to plan recreational events. Farmers need to know whether the weather will remain fine long enough for them to plow their fields or bring in the harvest.

When you travel by air, the pilot and air traffic controllers have detailed knowledge of the weather conditions along the route you will take. This tells them the height at which it is best to fly, whether they will need to go around large storms that could give you an uncomfortable or unsafe ride, and whether the destination airfield will be clear of fog, low clouds, runway ice, or strong winds. Lives may depend on this information.

Similarly, fishermen, who trust their lives as well as their livelihoods to small boats, need to know whether it is safe to go out to sea. Even a large ship can be damaged, and in extreme cases sunk, if it sails into a hurricane. Sea captains need to know whether there are any severe storms along the routes they plan to follow.

Weather can even have political significance. During the 1996 presidential election campaign, bad weather forced candidate Bob Dole to abandon his election tour of New Jersey. A morning that started misty turned savage when Tropical Storm Josephine began pouring torrential rain on both the contender and his outdoor audience. Did a storm influence the outcome of the election? Probably not, but who can say?

As you know only too well, the weather changes constantly. It changes more often and more dramatically in some parts of the world than in others, but even in the humid tropics it is not always the same. You might think that in a tropical rain forest it rains every day, but it does not. Sometimes there are dry spells and the plants are short of water. In a desert it rarely rains, but there are no places where it never rains or snows at all; when the desert rains do come, they usually do so in torrents. There are towns in some deserts, and occasionally they are flooded.

It is this changeability that makes weather forecasting so difficult. If the changes were regular, like the phases of the Moon or the movement of the Sun across the sky, we would hardly need weather forecasters at all. We could look up tomorrow's weather, or next month's, in a book.

These are only the short-term changes, the changes that take place from one day to the next. Weather also changes on much longer time scales. Weather itself has a history.

If you could travel back in time 10,000 years, you would find a world in which some people were just beginning to domesticate animals and plant crops, but most still obtained their food, clothing, fuel, and timber for building by gathering wild plants. Their meat was won by fishing, gathering seafood, hunting game, and stealing meat from animals killed by more formidable hunters, such as big cats and wolves. In the world they inhabited, ice sheets were starting to retreat from a large part of North America, Europe, and Asia. The last ice age was coming to an end. In what are now Maryland and Virginia, the climate and landscape were like those you find today in northern Canada and Alaska; what is now the Great Salt Lake, in Utah, formed part of a huge inland sea called Lake Bonneville, which covered nearly 20,000 square miles (520,000 square kilometers) and was in places 1,000 feet (300 meters) deep. In those days you could have walked from North America across the ice sheet to Greenland, and from there to Europe. The North Sea was mostly dry land; Britain was still joined to the European mainland across the Dover Strait, and Alaska was joined to Siberia across the Bering Strait.

Land bridges existed because sea levels were low, due to the large amount of water that formed the ice sheets. You can still find evidence of those past sea levels. There are places on some coasts where the remains of forests can still be seen offshore. At very low tides they are revealed, not as trees, of course, after all this time, but recognizably as wood. Many of what are now estuaries were once river valleys, drowned as the rising sea flooded inland.

At other times sea levels were higher, suggesting that the climate has been so much warmer in the past that the polar ice sheets were much smaller than they are now. You can find evidence of this, too, along some coasts, in the form of "raised beaches." These are layers of ground well above the highest level reached by modern tides, containing seashells that must once have lain on the shore.

Still further back in time, beyond the ice ages, there is evidence of climates even more dramatically different. Coal is found in many parts of the world. Perhaps there are, or once were, coal mines not far from where you live. Coal is made from partly decayed vegetation, and it forms only in swamps along tropical coasts. What are now coalfields were once tropical coastal swamps. If you live far inland, but the soil is very sandy, your neighborhood may once have been part of a dry, sandy desert.

Big differences such as these are due to more than the changing climate, of course. All the continents of the world are moving. North America and Europe are growing farther apart, for example, as the

Atlantic Ocean widens. It is a slow process. The Atlantic is widening by about 1 inch (2.54 centimeters) a year. Over many millions of years this adds up to a large distance, however, and at the time our coalfields formed, the climate was tropical because the lands in which they are found lay in the tropics.

This was long ago, but climates continue to change. After the last ice age ended, the average temperature over much of the world was about 2–5°F (1–2.8°C) warmer than it is now. There are caves in the central Sahara Desert containing drawings and paintings made by people who lived up to 8,000 years ago. What is interesting about them is the way of life they depict. One shows a hippopotamus being hunted by men in canoes. Another shows people driving a herd of cattle. Clearly, what is now the biggest desert in the world was then a much more pleasant place to be, at least in some places, and there is other evidence. Fish bones and old shorelines show that Lake Chad was once a big inland sea. Until around 5,000 years ago, most of the Sahara received enough rain to support grassland vegetation, and it was crossed by rivers that seldom ran dry.

During the whole of this long period of warm, moist weather, average temperatures over Europe and probably over most of the world were about 4°F (2.2°C) higher than they are now. There is a pond tortoise (*Emys orbicularis*) that lives in Europe only in places where the average July temperature is at least 65°F (18.32°C). Until about 3000 B.C. it lived in Denmark but now the pond tortoise finds its habitat in Mediterranean countries.

Already, though, the climate was becoming drier. Some historians believe the rise of civilizations in Egypt and the valleys of the Euphrates and Tigris rivers was stimulated by the need to organize more efficient food production as the rains became lighter and less reliable and large areas of land gradually turned into desert. It was not only the Middle East that was affected. A similar drying of the climate occurred in the Indus Valley, in northern India, and in China.

Since then, the climate has been more variable, but the final years of the warm period coincided with great advances in civilizations throughout the world. Forests and grasslands extended much farther north than they do now, and people lived in regions that are now barely habitable.

The variability of more recent climates involves fluctuations measured over centuries, marked by temperature changes of only a few degrees. These variations seem small, but they mean markedly cooler or warmer summers and longer or shorter winters, which have a big effect on the way people live. Pictures of the people who lived in Mycenae, around 1,500 B.C. show them naked or very lightly dressed. Some centuries later, the early Greeks wore warm clothing, and there are reports of the Tiber River freezing at Rome around 300 B.C. and snow and ice lasting for some time. Yet by the first century A.D. the Italian climate was warm, and snow was a very rare sight in Rome. When the Romans occupied Britain, they introduced the growing of vineyards for wine; by about 300 A.D. British vineyards may have produced sufficiently to satisfy domestic demand, for about then records of wine imports cease.

After about 400 A.D. the climate changed for the worse, becoming cooler and stormier. In the winter of 859–860, the sea froze along the coast near Venice, and the ice was thick enough to bear the weight of loaded wagons; in 1010–11, even the Nile froze. Farther north, however, the warm conditions continued much longer. Around 980, Norse settlers led by Erik the Red founded colonies in Greenland. Eventually there were about 3,000 people living there on nearly 300 farms. That does not necessarily mean the weather was especially warm, but another story does. A cousin of Erik's, called Thorkel Farserk, needed a sheep for the dinner he planned for his cousin. The sheep was on an island more than 2 miles (3.2 kilometers) away and no boat was available, so Thorkel swam out and carried the sheep back. If the story is true, Thorkel would have died if the sea temperature had been lower than 50°F (10°C). Today the sea in those parts is never warmer than 43°F (6°C).

Cereals were being grown in Norway up to about 1000, but within a century cooler weather was reducing the farmed area. Warm conditions lasted until around 1300 farther south, however, and the English were producing high-quality wine, with summer temperatures up to 2°F (1°C) higher than those of today; it was warmer still in central Europe.

In North America, the period from about 700 to 1200 was relatively warm and moist. Grasslands were being replaced by forests, and people were farming along the river valleys into the Great Plains. By about 1150 almost every habitable area was occupied. Some of the inhabitants lived in towns with stone buildings several stories high; in this period the cliff homes of the Mesa Verde were built. People made roads, signaling stations, and channels to carry water to their homes and fields. That was the peak, however, and as the climate became drier people crowded along river banks. In time, the farms and towns were abandoned. Few remained after 1300.

All over the world, climates were cooling. It was a gradual change that began around 1300 and continued until the weather reached its coldest, at the end of the 17th century. Between 1690 and 1699, average temperatures in England were 2.7°F cooler than they are now. This was the Little Ice Age. Alpine glaciers advanced; European settlers, just starting to arrive in North America, encountered severe weather. There was ice around the edges of Lake Superior in June 1608, and land in the north that is now free of snow in summer remained covered by snow and ice throughout the year.

We are still living with some of the consequences of the Little Ice Age. It brought such harsh conditions in Scotland that famines and food shortages were common, triggering the start of an emigration of Scots people to many parts of the world; many young men found work as professional soldiers. It was said that in 1700 there was hardly an army in Europe that did not have some Scottish officers. In 1612, the Scottish king James VI, who had become James I of England following the union of the two crowns, giving him powers in Ireland, evicted Irish farmers from the more fertile and sheltered lands of Ulster and "planted" Scottish farmers in their place. This helped relieve the hardship in Scotland while increasing the king's control of the Irish. By the end of the 17th century, the Scottish population of Northern Ireland had reached 100,000 and was still rising rapidly.

The Little Ice Age eased slowly, and there were warm years as well as cold ones, but it did not really end until the latter part of the 19th century, when temperatures started to rise once more. It is possible that the climatic warming scientists believe has occurred since then is part of this recovery.

Most climatologists believe the present climatic change is being accelerated by our release of certain gases into the atmosphere. This, they propose, will lead to significant global warming due to an enhanced greenhouse effect.

Enhanced greenhouse warming occurs because although all atmospheric gases are transparent to incoming, short-wave, solar radiation, some partially absorb the long-wave heat radiated from the land and sea surface after it has been warmed by the Sun; the absorbed heat warms the air. The most important "greenhouse gases" are carbon dioxide and nitrous oxide, released by burning carbon-based fuels; methane, released by bacteria in the digestive systems of cattle and in the mud of rice paddies; and chlorofluoro-carbon (CFC) compounds and their substitutes, which we use in refrigerators, freezers, and air conditioners, and in making foam plastics. If scientists are right, and we continue to release these gases, by the year 2100 the average temperature may rise about 4.5°F (2.5°C) above its present value.

Warming by this amount would give us climates as warm as those of 8,000 years ago. There have been warm periods since then, but none so warm as is now being predicted, and no warming has ever happened so fast. It is still too soon to know whether the prediction is accurate or, indeed, to be certain that our release of greenhouse gases is affecting the climate at all. Nevertheless, we would be foolish to ignore the warning; although it may turn out to be invalid, it is by no means impossible.

Today may be fine and warm. Yesterday may have been cool and wet. Weather changes from day to day and from season to season, but it also changes over decades and throughout the centuries. In the past people have lived with weather markedly different from the weather we experience now. The story of our climate is fascinating; paleoclimatologists, the scientists who use evidence they find in pollen grains, tree rings, and traces of chemicals stored deep below the ice sheets to reconstruct what past climates were like, are gradually unraveling it.

Climate has a history that, like human history, is still continuing. Little by little we are able to tell its story as it unfolds, but the tale has not yet come to an end.

History of the science of weather

People have always been interested in the weather. This is not surprising. Crops grow or wither according to the sunshine and rain they receive, and the weather determines the amount and condition of the pasture on which farmers graze their animals. Even today, in many parts of the world bad weather can mean a failed harvest and famine.

Violent weather has always brought more immediate disaster. Prolonged, heavy rain has caused floods in which thousands have died. Severe hailstorms have flattened crops, hurricanes and tornadoes have destroyed everything in their paths, and blizzards have not only buried and frozen people, they have done the same to the livestock on which the survivors would have depended for food. Such catastrophes have occurred throughout history, and until very recently they have unleashed their violence without warning.

It is small wonder that in ancient times most people believed the weather was controlled by gods who used it to reward or punish humans. Little provocation was needed to make bad-tempered gods start hurling thunderbolts in all directions, yet in a better mood those same gods could send rain to make crops grow and sunshine to ripen them. In the Old Testament, God "commandeth, and raiseth the stormy wind" and "maketh the storm a calm" (Psalm 107).

By the 5th century B.C., however, the idea was developing that the weather is produced by natural, not supernatural, forces. Aristotle (384–322 B.C.) wrote on many topics, and his *Meteorologica*, written in about 340 B.C. is the earliest known book on the scientific study of weather. It also gives us the word *meteorology* (literally, "discourse on lofty matters"). Today, a meteorologist is a scientist who studies weather, the conditions we experience day by day. Climate is the kind of weather that occurs over a very large area, such as a continent or even the entire world, over a long period. Scientists who study climate are called "climatologists." Some climatologists, called "paleoclimatologists," specialize in the reconstruction of climates in prehistoric times.

Although Aristotle was a keen observer, neither he nor his followers had the instruments with which to make accurate measurements, and they performed no experiments. They simply tried to find natural explanations for what they saw. Other writers described particular weather conditions, especially winds, and their effects. The Roman writer Pliny (*c.* 23–79), for example, wrote that the coldest winds are from the north, that southerly and southwesterly winds are damp in Italy, and that northwesterly and southeasterly winds are usually dry.

This kind of "weather lore" allows people to predict the weather by noting present conditions and recalling what has usually followed them in the past. We still use it when we predict, often correctly, that a red sky at sunset will be followed by fine weather and a red sky at dawn by bad weather. Weather lore extended much further, however, into predictions based on observations of plants and animals. Some are reliable, because many plants and insects are sensitive to changes in humidity, but others are not. It is hard to see why a cock crowing at sunset should herald rain by dawn, but this used to be believed in parts of England.

Until suitable instruments were invented, this kind of forecasting was the most meteorologists could achieve. A vast amount of observations were noted, sometimes mixed with superstition and astrological interpretations, but the scientific study of weather made no significant advance from the time of Aristotle until the 17th century.

In 1643, Evangelista Torricelli (1608–47) tried to discover why it is impossible to pump water from a well more than 33 feet (9.9 meters) deep. He thought the air might have weight and press down too strongly. To test this, he sealed a glass tube at one end, filled it with mercury, then inverted it with the open end in a dish of mercury. He found the mercury fell to a height of about 30 inches (76.2 centimeters). Later, he noticed that the height of the mercury column varied when the weather changed. Torricelli had invented the barometer, and it was not long before changes in the height of the mercury were linked to changing weather. Some years later, Robert Hooke (1635–1703), the English physicist and instrument maker, equipped barometers with a scale reading "change," "rain," "much rain," "stormy," "fair," "set fair," and "very dry," which is used to this day on many household barometers. The hygrometer, which measures humidity, was invented in 1687 by the French physicist Guillaume Amontons (1663–1705); the hair hygrometer, widely used today, was invented by the Swiss physicist Horace de Saussure (1740–99) in 1783.

Galileo (1564–1642) invented a thermometer based on the expansion of air, but because the volume of a mass of air varies according to atmospheric pressure, it was very inaccurate. The first mercury thermometer is believed to have been invented in 1714 by Gabriel Fahrenheit (1686–1736).

Together with the rain gauge, which had been used since ancient times, the barometer, hygrometer, and thermometer allowed meteorologists to make accurate atmospheric measurements. Scientific understanding of atmospheric processes increased steadily, but it was still impossible to immediately study the conditions occurring simultaneously at places hundreds of miles apart. The reason is obvious: In those days no message could travel faster than the speed of a galloping horse. By the time observations taken many miles apart had been collected at a central point, it would be several days after the recorded events.

Efficient collection of records became possible in 1844, the year Samuel Morse (1791–1872) persuaded the U.S. Congress to finance the construction of a telegraph line between Baltimore and Washington, D.C. Morse invented the telegraph and the binary code it used, and it was successful. Within 20 years France, followed by the United States, Britain, and other countries, had established weather stations reporting by telegraph to centers for analysis and forecasting. Modern meteorology may be said to have begun around the middle of the 19th century.

Although the weather experienced at a particular place may seem to be very localized, the weather system producing it is likely to extend over hundreds of square miles and to a height of several miles. Meteorologists, who study weather systems and their movement and use the information they gather to forecast the weather, now rely on a worldwide network of stations that record and forward measurements every few hours.

Surface weather stations form a network covering the world, although not very evenly. They measure wind strength and direction, temperature, humidity, air pressure, cloud type and amount, and visibility. Some take measurements each hour, just before the hour; others manage to take measurements only every six hours. To avoid the confusion that might result if weather stations in different time zones reported in their own local time, all stations, no matter where they are in the world, report their measurements in Greenwich Mean Time, which is the time at London, England.

Some stations measure conditions at ground level; others sample the upper atmosphere, using balloons called "radiosondes." They are called radiosondes because they take "soundings" (measurements) and transmit them to receiving stations by radio. Radiosondes measure atmospheric conditions above ground level and to a height of about 80,000 feet (24,000 meters), in the middle of the stratosphere. Balloons were first used for this purpose in 1927. To make sure that measurements from around the world can be combined to produce a comprehensive picture of atmospheric conditions at a particular time, every weather station, no matter where it is, releases one radiosonde every day at midnight and a second at noon Greenwich Mean Time. The National Meteorological Center, in Washington, D.C., receives about 2,500 sets of radiosonde data every day.

The balloon itself is spherical, about 5 feet (1.5 meters) in diameter, and filled with hydrogen. Beneath it is a cable nearly 100 feet (30 meters) long with a package of instruments attached to its lower end; the cable needs to be this long to make sure air movements around the balloon do not interfere with instrument readings. The standard instruments package comprises a very sensitive thermometer, a hygrometer to measure humidity, and a barometer to measure air pressure. There are also timers, switches to turn the instruments on and off at predetermined times, a radio transmitter, batteries to supply power, and a parachute to return the instruments safely to the ground.

After the radiosonde is released, it climbs steadily, at about 15 feet (4.5 meters) per second. As it rises, its hydrogen expands, and when the radiosonde reaches a height of around 80,000 feet (24,000 meters), the balloon bursts and its instruments parachute to the ground, where they are recovered and returned to the weather station from which they were launched. During its flight, the radiosonde broadcasts its measurements to the ground station.

In addition to its instruments, the radiosonde carries a radar reflector immediately below the balloon. This strongly reflects radar pulses and allows the movement of the radiosonde to be tracked from the ground. Before radar was invented, balloons were tracked visually, but, of course, they disappeared from view as soon as they entered clouds.

As the radiosonde ascends, it moves horizontally with the wind, which usually changes direction and speed in different layers of air. Balloons that are tracked by the signals received from their radar reflectors in order to study winds at very great heights are sometimes called "rawinsondes" (for radar **wind**-sounding). Such tracking provides accurate information on the wind speed and direction in each atmospheric level through which the device passes. In years to come, rawinsondes will probably broadcast their precise locations using the Global Positioning System, in which orbiting satellites pinpoint the source of a radio signal transmitted to them.

Since the first weather satellite (Tiros 1) was launched in 1960, satellites have played an increasingly important part in tracking weather systems. They transmit to ground stations photographs taken in one or more visible light wavelengths or in the infrared. There are two types of weather satellites: Those in geostationary orbits remain in a fixed position relative to the Earth's surface. Those in sun-synchronous orbits remain in a fixed position relative to the Sun. These orbits carry them around the Earth about every 102 minutes, passing close to the North and South Poles.

Information from satellites and surface stations is sent to centers where it is fed into supercomputers for processing. The data from these supercomputers are used to compile weather maps from which the maps appearing in newspapers and on television are derived.

The World Meteorological Organization is the United Nations agency responsible for coordinating meteorological and climatological studies throughout the world. In the United States, the National Oceanic and Atmospheric Administration (NOAA), part of the U.S. Department of Commerce, operates the National Weather Service. In Britain, the Meteorological Office produces and issues weather forecasts and other relevant information.

The weather forecasts we receive daily are only part of the service these organizations provide. They also produce specialized forecasts for ships, airlines, fishermen, farmers, climbers, and others.

A chronicle of destruction

5,000 Years of Dangerous Weather

c. 3200 B.C.

The Euphrates River flooded, inundating the city of Ur (in modern Iraq) and the surrounding countryside. In 1929, archaeologists discovered evidence of the flood, in the form of a layer of flood deposits 8 feet (2.4 meters) thick.

c. 2200 B.C.

Drought, with intermittent fierce storms, caused deserts to spread and harvests to fail over large parts of the Mediterranean region and the Near and Middle East. This caused the collapse of the empire of Akkad, the Old Kingdom of Egypt (during which the greatest of the Pyramids were built), several Bronze Age cities in Palestine, civilizations in Crete and Greece, and the cities of Mohenjo-Daro and Harappa in the Indus Valley. Evidence for the drought is provided by dust deposits, falling water levels in lakes, and ocean sediments.

A.D. 245

Floods inundated thousands of acres in Lincolnshire, England.

c. 300

Drought in central Asia coincided with local wars that led nomads to invade northern China, causing the collapse of the Tsin dynasty. Refugees fleeing south stimulated cultural development in southern China, Korea, and Japan.

c. 520

The whole of Cantref y Gwaelod, a country bordering Cardigan Bay in west Wales, was inundated when a severe storm breached a dike.

678

A drought began in England and lasted three years.

1099

A storm surge along the coasts of England and the Netherlands caused 100,000 deaths.

1103

Severe gales on August 10 caused severe damage to crops in England.

1140

A tornado caused extensive damage in Wellesbourne, Warwickshire, England.

1246

Severe drought began in what is now the southwestern United States and lasted until 1305, being most intense from 1276 to 1299.

1276

Drought began in England and lasted until 1278.

1281

A typhoon from the China Sea destroyed most of a fleet of Korean ships carrying a Mongol army on its way to invade Japan. The wind, which saved the Japanese from foreign conquest, was believed to be a "divine wind" (*kamikaze*) sent by the gods.

1305

Drought in England caused the hay crop to fail, and many farm animals died. Mortality was high among humans, due partly to a smallpox epidemic.

1333

The Arno River, Italy, overflowed on November 4, causing 300 deaths.

1353

Drought in England, lasting from March to the end of July, caused some famine. The following year was also dry.

1421

An estimated 100,000 people drowned on April 17 when the sea broke through dikes at Dort, Netherlands.

1558

A tornado struck the area around Nottingham, England, on July 7. All houses and churches within a mile of the city were destroyed, a child was lifted to 100 feet (30 meters) and dropped, trees were thrown more than 200 feet (60 meters) and five or six people were killed.

1592

On December 16–17, the river Niger overflowed, flooding the city of Timbuktu, capital of the Kingdoms of Mali and Songhay, and causing the population to flee. The floods were caused by unusually heavy rains in Guinea, near the source of the river.

1638

On Sunday, October 21, a tornado accompanied by ball lightning struck the church at Widecombe-in-the-Moor, on Dartmoor in southwest England, during the morning service. Different accounts recorded between 5 and 50 people killed.

1642
The Yellow River, China, flooded when dikes were deliberately broken during a peasant revolt led by Li Tzu-cheng in the hope of inundating and capturing the besieged city of Kaifeng. About 900,000 people were killed.

1654
Severe drought in the Midi of France, lasting several years, led people at Périgueux to visit the shrine of St. Sabina and ask her to intervene and bring rain.

1666
Drought in England reduced the flow of water in the river Thames so seriously it threatened to ruin boatmen. By September wooden structures in London were so dry a small spark could ignite them (that is when the Great Fire of London occurred).

1674
A blizzard along the border between England and Scotland began on March 8 and lasted 13 days.

A severe gale on December 21 uprooted entire forests in Scotland.

1703
Hurricane-force winds along the English Channel swept the southern coast of England on November 26 and 27, destroying 14,000 homes and killing 8,000 people. Eddystone Lighthouse (off Plymouth) was destroyed, and 12 warships sank. Damage in London was estimated at 2 million pounds.

1726
March floods in southern England raised the level of the Thames and inundated Salisbury Cathedral.

1730
Drought began in England and lasted until June 1734.

1740
A hurricane struck London on November 1.

1757
On November 12 the river Liffey overflowed, causing extensive damage in Dublin.

1762
A blizzard in England lasted 18 days in February, killing nearly 50 people.

1780

A prolonged Caribbean storm, known simply as "the Great Hurricane," swept through the West Indies in October. The British fleet, anchored at Barbados, suffered severe damage, as did a Spanish fleet in the Gulf of Mexico, nearly 2,000 miles (3,200 kilometers) distant.

1824

Ice jamming the river Neva, Russia, caused flooding in St. Petersburg and Kronshtadt in which 10,000 people drowned.

1829

In Morayshire, Scotland, on August 3, 4, 27, and 29, there was severe flooding known as the "Moray Floods," when the rivers Spey and Findhorn overflowed. Many lives were lost.

1831

A hurricane struck Barbados, killing 1,477 people and causing extensive damage to buildings and ships moored in the harbor.

1852

Heavy rains brought widespread flooding in central England in September, turning the Severn Valley into a continuous "sea."

1853

In summer, heavy rain brought widespread flooding in England, destroying crops and killing many sheep.

1854

During the Crimean War, a hurricane on November 14 devastated the British fleet, moored off Sevastopol because the Russians had mined the harbor. Winter supplies for the army were lost, resulting in severe deprivation among the troops.

1865

In June, a tornado moved through Viroqua, Wisconsin, demolishing 80 buildings and causing more than 20 deaths.

1875

The river Thames rose, some said by more than 28 feet (8.5 meters), on November 15, causing widespread flooding in London.

1876

The Bakarganj cyclone formed in the Bay of Bengal and moved north. Coinciding with the high monsoon river level of the Ganges, it caused flooding in islands in the Ganges Delta and on the mainland in which 100,000 people drowned in half an hour.

1879

On December 28, the Tay Bridge, carrying the rail line across the Firth of Forth, was struck by two tornadoes simultaneously just as the evening mail train was crossing on its way from Edinburgh to Dundee. The bridge was destroyed and the train fell into the river, far below. The number of passengers is uncertain, but between 75 and 90 people lost their lives.

1881

A typhoon in China on October 8 killed an estimated 300,000 people.

1887

The Yellow River, China, burst its banks in September and October, causing extensive flooding in Honan, Shantung, and Hebei (Chihli) provinces, beginning with a breach in the dikes near Cheng-chou. Flood water flowed through 1,500 or more towns and villages and covered an area of about 10,000 square miles (26,000 square kilometers). Estimates of the number of fatalities ranged from 900,000 to 2.5 million.

1888

From January 11 to 13, Montana, the Dakotas, and Minnesota experienced the most severe blizzard these states had ever known.

Following a mild winter, a blizzard with winds gusting to 70 MPH (112 KPH) struck the eastern United States, from Chesapeake Bay to Maine, from March 11 to 14. Temperatures fell close to 0°F (-17.8°C), an average of 40 inches (101.6 centimeters) of snow fell over southeastern New York and southern New England, and more than 400 people died, 200 of them in New York City.

On March 28 a storm and storm surge caused great damage in Wellington, New Zealand.

1891

On February 7 severe blizzards, continuing for several days, caused many deaths in Nebraska, South Dakota, and other U.S. states.

Between March 9 and 13, more than 60 people died in blizzards in southern England, and near-hurricane-force winds destroyed about 14 ships in the English Channel, with heavy loss of life.

1892

On January 6 a cyclone caused many deaths in Georgia and neighboring states.

1894

The river Thames overflowed on November 15 between Oxford and Windsor, England, causing extensive flooding and severe damage.

1900

A category 4 hurricane struck Galveston, Texas on September 8, killing 6,000 people, injuring more than 5,000, and destroying half the buildings in the town.

1903

On June 14 a thunderstorm, lasting from 4:00 P.M. to 5:00 P.M., brought heavy rain to the Willow Creek basin in the Blue Mountains foothills, sending a flash flood through Heppner, Oregon (pop. 1,400), and killing 247 people.

1919

A category 4 hurricane struck the Florida Keys, killing 900 people.

1922

On January 15 a hurricane with torrential rain struck the area around Tamanrasset, Algeria. The rain continued into the following day, floods sweeping away huts and gardens beside a wadi (a dry channel in a desert, through which a river flows very occasionally) and destroying the wall of the Père de Foucauld fort. The collapsing wall buried 22 people, killing 8 and injuring 8 more.

1925

A series of tornadoes, possibly seven in all, developed over Missouri in March and crossed Illinois and Indiana, traveling 437 miles (699.2 kilometers)—one moving more than 120 miles (192 kilometers) in three hours. Altogether, 689 people were killed. On March 18, a tornado struck Annapolis, Maryland overturned passenger trains and lifted 50 motor cars, carried them over rooftops, and dropped them in the fields beyond.

1927

The Mississippi River flooded in April, following heavy rain that had begun in August 1926. High water levels in the main river caused backwater to raise the levels of several tributaries, which also flooded. Eventually flood waters covered more than 25,000 square miles (65,000 square kilometers) in seven states. In places they were 18 feet (5.4 meters) deep and 80 miles (128 kilometers) across. The worst affected states were Louisiana, Arkansas, and Mississippi. The floods finally receded in July. Officially, the death toll was given as 246, but it may have been much higher, and nearly 650,000 people lost their homes.

1928
A category 4 hurricane struck Lake Okeechobee, Florida, causing floods that killed, 1,836 people.

1930s
Drought over most of North America was most severe in the Great Plains, especially in Kansas and the Dakotas. Between 1933 and 1935 soil, exposed by plowing and reduced to fine dust by the weather, began to blow away in a series of dust storms. The most severe storms, in 1934 and 1935, carried soil to the Atlantic coast, and erosion affected 80 percent of the Great Plains. After this the most severely drought-stricken area became known as the "Dust Bowl." About 150,000 people migrated from the region. Earlier droughts had occurred in the Great Plains between 1825 and 1865; 1881, 1894–95, 1910, 1936, 1939–40; and later in the 1950s and 1960s.

1931
The Yellow River, China, overflowed between July and November, flooding about 34,000 square miles (88,000 square kilometers). An estimated one million people died from drowning and in the famine and epidemics that followed, and 80 million were made homeless.

The Yangtze River, also in China, rose 97 feet (29.6 meters) after heavy rain. More than 3.7 million people died, most from the subsequent famine, and the cost of damage was estimated at $1.4 billion.

1935
A category 5 hurricane struck the Florida Keys on Labor Day, generating winds of 150–200 MPH (240–320 KPH) and killing 408 people.

1937
The Mississippi River flooded in January, after heavy rains in the Ohio River basin caused the Ohio to flood, discharging a huge amount into the Mississippi and raising the level 63 feet (19 meters) at Cairo, Illinois. Water flowed back into the tributaries of both rivers. The flood waters covered about 12,500 square miles (32,400 square kilometers), destroyed 13,000 homes, and 137 people died. Damage was estimated at about $418 million.

1938
A category 3 hurricane struck New England, killing 600 people.

The Yellow River, China, flooded on June 9 when Kuomintang forces broke the dikes, hoping to cause a small, local flood that would check a Japanese advance on Cheng-chou. The river flowed out of control, submerging more than 9,000 square miles (23,300 square kilometers). The flood caused 500,000 deaths, and six million people lost their homes.

1944

A U.S. naval fleet in the Philippine Sea inadvertently sailed directly into Typhoon Cobra, with winds of up to 130 MPH (208 KPH) and waves up to 70 feet (21 meters) high. Altogether 3 destroyers sank, 150 carrier-borne aircraft were destroyed, and 790 sailors were killed.

1952

On August 15 in Exmoor, southwest England, following heavy rain on already saturated land—amounting to 9 inches (2.7 meters) in 24 hours—the West and East Lyn rivers overflowed. The rivers cut a new channel and inundated the coastal village of Lynmouth, depositing about 200,000 tons (220,000 tonnes) of debris, including large boulders, on the shore. In all, 34 people died, 93 homes were destroyed, and repairing the damage cost an estimated $2 million.

The Nishnabotna River, a tributary of the Missouri, overflowed inundating more than 66,000 acres (26,700 hectares) of farmland.

1953

A storm surge caused nearly 2,000 deaths along the coasts of England, the Netherlands, and Belgium.

A typhoon caused extensive damage in Nagoya, Japan. One million people were homeless, and 100 died.

1954

A typhoon struck the island of Hokkaido, Japan, killing 1,600 people.

Hurricane Hazel, formed in the Lesser Antilles, struck Haiti on October 12, killing an estimated 1,000 people. It reached Myrtle Beach, South Carolina, on October 15, devastating coastal towns, producing a 17-foot (5.1-meter) storm surge, and killing 19 people. It then crossed North Carolina, Virginia, Maryland, Pennsylvania, and New York State, killing 76 people. Winds of more than 100 MPH (62.5 KPH) were recorded at LaGuardia and Newark airports in New York and New Jersey. Total damage in the United States amounted to $250 million. Hazel continued into Canada, killing 80 people, leaving 4,000 homeless, and causing $100 million of damage. On October 18, Hazel moved into the Atlantic, eventually producing heavy rain and strong winds in Scandinavia.

The Yangtze River, China, overflowed. About 30,000 people died, most from the subsequent famine.

Severe thunderstorms on August 17 filled dry creeks and ravines at Farahzad, Iran, sending a wall of water 90 feet (27 meters) high crashing through a shrine where 3,000 people were worshipping. An official shouted a warning, but more than 1,000 people died.

On December 8, a severe tornado struck west London, England, during the afternoon rush hour. It made a track 100–400 yards (90–360 meters) wide extending 9 miles (14.4 kilometers) through Chiswick, Gunnersbury, Acton, Golders Green, and Southgate, causing extensive damage.

1955

During the day and night of August 18 almost 8 inches (12.8 centimeters) of rain fell on land drained by the Quinebaug River, Connecticut. The land was already saturated by previous intense rain, and on the morning of August 19 a series of dams broke, releasing flash floods on the town of Putnam (pop. 8,200), to the south. Roads, bridges, railroad embankments, and one-quarter of all the buildings in Putnam were destroyed. Water entered a warehouse containing 20 tons (22 tonnes) of magnesium, which ignited in a series of explosions. The waters receded a week later. Damage was estimated at $13 million, but thanks to rapid evacuation and efficient emergency services, no lives were lost.

Hurricanes brought rains to the eastern United States that caused flooding in southern New England, southeastern New York, eastern Pennsylvania, and New Jersey. Damage was estimated to cost $686 million, and 180 people died.

The Macquarie, Castlereagh, Namoi, Hunter, and Gwydir rivers, in New South Wales, Australia, flooded. Nearly 40,000 people were made homeless, and 50 died.

1956

Extremely heavy rains in Australia caused rivers to flood, forming a continuous "sea" 40 miles (64 kilometers) wide between the towns of Hay and Balranald.

Hurricane Audrey, category 3, reached the U.S. Gulf coast near the Louisiana–Texas border on June 27, with winds of 100 MPH (160 KPH) and a 12-foot (3.6-meter) storm surge. Nearly 400 people died.

1957

Hurricane Diane brought heavy rain to the United States in August, a few days after Hurricane Connie had saturated the ground. More than 190 people died, and damage cost $1.6 billion.

1959

Typhoon Vera struck the island of Honshu, Japan, in September, destroying 40,000 homes. It killed nearly 4,500 people and left 1.5 million homeless.

A hurricane killed 2,000 people on the west coast of Mexico.

A cyclone from the Bay of Bengal left 100,000 people homeless on the islands of the Ganges Delta and the adjacent mainland.

1961

Typhoon Muroto II caused a 13-foot (3.9-meter) storm surge and flooding in Osaka, Japan, in September, killing 32 people.

In September the most powerful hurricane ever experienced in the region struck Galveston, Texas. There was extensive wind damage and flooding, but the sea wall that had been erected in the aftermath of the 1900 hurricane held back the worst of the storm surge waves, and less than 50 people died.

1963

Rocks and scree (a large pile of loose rocks) on the slopes of Mount Toc, Italy, destabilized and caused a full reservoir, behind an 860-foot (258-meter) dam at the junction of the Vaiont and Piave rivers, to move calamitously after heavy rains throughout the summer. On October 9 at 10:41 P.M., 314 million cubic yards (240 million cubic meters) of rock fell into the reservoir, causing a wave 330 feet (101 meters) higher than the dam wall to release a wall of water that was 230 feet (70 meters) high when it reached the town of Longarone, 1 mile (1.6 kilometers) downstream. It killed almost all the inhabitants, then flooded the villages of Pirago, Villanova, and Rivalta. In all, 2,600 people died.

1964

On March 27, earthquakes in Alaska caused tsunamis along the U.S. Pacific coast.

1966

The Arno River, Italy, overflowed on the night of November 3 after prolonged heavy rain, flooding Florence, in places to a depth of 20 feet (6 meters). There was extensive damage to historic buildings and art treasures, 35 people died, and 5,000 were made homeless.

1968

Chicago was brought to a standstill for several days when 24 inches (60.96 centimeters) of snow fell, driven by winds up to 50 MPH (80 KPH).

1969

Hurricane Camille, category 5, struck Mississippi and Louisiana on August 17 and 18. It killed about 250 people on the coast and caused $1.42 billion of damage. As Camille weakened and headed south, then east, over the Blue Ridge Mountains, Virginia, its winds gathered moist Atlantic air and funneled it through the narrow valleys of the Rockfish and Tye rivers. There it encountered an advancing cold front, with thunderstorms. The resulting 18 inches (45.72 centimeters) of rain, starting at 9:30 P.M., flooded 471 square miles (1,220 square kilometers) in Nelson County. Flash floods destroyed or damaged 185 miles (298 kilometers) of road and deposited mud and debris, in some places to a depth of 30 feet (9 meters); 125 people died by drowning or by being crushed by boulders. The floods subsided after 80 days.

On October 24, severe flooding in Tunisia left more than 300 people dead and 150,000 homeless.

1970

In November, one of the worst natural disasters of the 20th century occurred when a cyclone moved north through the Bay of Bengal and devastated Bangladesh, killing about 500,000 people.

1972

Hurricane Agnes, category 1, struck Florida and New England, causing $2.1 billion of damage.

Torrential rain in the Black Hills caused extensive flooding throughout South Dakota on June 10. Hundreds of people died.

Seasonal rains failed in the Ethiopian highlands from June to September, and crops could not be planted in the worst affected provinces. This led to the loss of 80 percent of the cattle and many camels in those provinces. In September 1973, it was estimated 100,000 to 150,000 people had died.

1973

On January 10, a tornado with 100-MPH (160-KPH) winds, lasting three minutes, destroyed buildings, killed 60 people, and injured more than 300 in San Justo, Argentina.

Hurricane-force winds struck the coasts of Spain and Portugal on January 17. At least 19 people died.

Severe storms brought snow, sleet, and freezing rain to the southeastern United States from February 8–11. On February 10, Macon, Georgia, had 16.5 inches (41.9 centimeters) of snow and ended with a total of 23 inches (58.42 centimeters), and the storms brought an average of 16 inches (40.6 centimeters) to parts of Georgia and the Carolinas. Charleston, South Carolina, had 7.1 inches (18 centimeters). Communications were disrupted, traffic halted, and crops killed.

The Caratinga River, Brazil flooded on March 26. Thousands of people were made homeless, and at least 20 died. Damage was estimated at $16 million.

Floods in late March and early April devastated towns and countryside in Tunisia. About 6,000 homes were destroyed, 10,000 cattle were killed, and about 90 people died.

A storm lasting 24 hours on April 9 generated huge waves on Lake Michigan. These pounded 28 miles (45 kilometers) of lakefront, causing damage estimated at $600,000. The storm killed 26 people.

Typhoon-force winds and rain struck Faridpur District, Bangladesh, on April 12, killing up to 200 people, injuring about 15,000, and leaving 10,000 homeless.

The Mississippi River and its tributaries flooded near St. Louis, Missouri, on April 29, submerging almost 1,000 square miles (2,600 square kilometers) and causing at least 16 deaths.

Tornadoes and rainstorms affected 11 states in the Southeastern United States on May 16–28. Alabama and Arkansas were declared disaster areas, and 48 people died.

Intense rain caused flooding on July 8 in the towns of San Pedro Itzican, Ocotlán, and Mazcola, on the shores of Lake Chapala, Mexico, killing at least 30 people.

Heavy rains on July 14–15 caused floods and resulting landslides in the Italian Riviera, killing 14 people.

After seven years of drought, Senegal, Mauritania, Mali, Upper Volta, Niger, and Chad suffered one of the worst famines of the century, alleviated by food and other supplies from many parts of the world. No estimate could be made of the number of deaths.

Unusually heavy monsoon rains along the Himalayas caused rivers to overflow in August in Pakistan, Bangladesh, and the Indian states of Uttar Pradesh, Assam, and Bihar. Entire towns were inundated; thousands of square miles of farm land were submerged; food stores, roads, railroads, and bridges were destroyed; and damage came to hundreds of millions of dollars. Thousands of people were killed and millions made homeless.

Prolonged rain in August caused floods in Irapuato, Mexico, that killed around 200 people and left 150,000 homeless. Damage was estimated at more than $100 million.

Heavy rains in October caused flash floods in states from Nebraska to Texas, killing at least 35 people.

Intense rain from October 19 to 21 caused flash floods in Granada, Almería, and Murcia, Spain. At least 500 people died, and damage was estimated at $400 million.

Typhoons struck Vietnam on November 10 and 11, bringing rains that destroyed bridges and buildings and ruined crops. At least 60 people died, and 150,000 had to leave their homes.

A typhoon and heavy monsoon rains caused flooding in many towns and along the Cagayan Valley, Philippines, from November 18 to 24, killing 54 people.

A cyclone struck the coast of Bangladesh on December 9, capsizing at least 200 fishing boats. Of the 1,000 people missing, many were feared drowned.

On December 13, flooding caused by heavy rain in Qafseh, Tunisia, killed 45 children.

On December 17 blizzards and low temperatures from Maine to Georgia, caused at least 20 deaths.

1974

January floods in eastern Australia caused 17 deaths and left about 1,000 people homeless.

In January, rains and tidal surges caused flooding in Situbondo, East Java, Indonesia, killing 19 people and leaving about 2,000 homeless.

There was flooding in eastern Australia in late January, caused by prolonged storms. In Queensland at least 15 people died, and damage was estimated at more than $100 million.

In February, heavy rain in northwestern Argentina caused flooding over half of Santiago del Estero Province and severe flooding in 10 other provinces. At least 100 people died, and more than 100,000 had to leave their homes.

Intense rainfall caused rivers to overflow in Natal, South Africa, during the night of February 10, sweeping away homes. More than 50 people died.

Torrential rains in late March, following months of drought, caused flooding in Tubarão, Brazil (pop. 70,000). The Tubarão River rose 36 feet (10.8 meters) in a few hours and overflowed, virtually destroying the city. It was estimated that between 1,000 and 1,500 people died and 60,000 were left homeless.

In early April, heavy rain caused severe flooding at Grande Kabylie, Algeria. At least 50 people died, and 30 were injured.

Thunderstorms and a total of 148 tornadoes developed on April 3, from Michigan to Alabama and Georgia; in 16 hours and 10 minutes this weather disaster traveled a combined distance of 2,598 miles (4,180 kilometers), eventually entering Canada, killing 323 people, injuring 6,000, and causing more than $600 million worth of damage. The small town of Guin, Alabama, was almost totally destroyed, but the most severe damage was in Xenia, Ohio, where nearly 3,000 buildings were damaged or destroyed and 34 people were killed.

Heavy rains in early May caused flooding in northeastern Brazil; about 200 people died.

A series of tornadoes crossed Oklahoma, Kansas, and Arkansas during the night of June 8. Altogether 24 people were killed.

Hurricane Dolores, accompanied by heavy rain, devastated an area housing poor people on the outskirts of Acapulco, Mexico, on June 17. At least 13 people died, 35 were injured, and 16 were unaccounted for.

A typhoon with heavy rain struck southern Japan on July 6 and 7, killing 33 people, injuring about 50, and leaving 15 unaccounted for.

Flooding in Bangladesh and northeastern India in August caused at least 900 deaths; an additional 1,500 people died in Bangladesh from a cholera epidemic that followed the floods. In Bangladesh, most of the grain crop was destroyed.

In August, monsoon rains caused extensive flooding in Luzon, Philippines. At least 94 people died, and more than one million needed disaster relief.

A cyclone struck West Bengal, India, on August 15, killing at least 20 people and injuring about 100.

Monsoon rains caused flooding along the Irrawaddy River, Burma, on August 20, inundating nearly 2,000 villages and washing away roads and railroads. At least 12 people died, and 500,000 were left homeless and stranded.

Hurricane Fifi, category 3, struck Honduras on September 20 with 130-MPH (209-KPH) winds and heavy rain. An estimated 5,000 people were killed and tens of thousands left homeless.

A small river flooded on November 19 in Silopi, Turkey, trapping members of a nomadic tribe camped nearby and killing 33 of them.

A 12-foot (3.6-meter) storm surge swept over low-lying islands in the southeastern Bay of Bengal on November 29 in the wake of a cyclone that had passed the previous day. The wave killed 20 people.

Cyclone Tracy destroyed 90 percent of the city of Darwin, Australia, on December 25, killing more than 50 people.

1975

In January, widespread flooding following heavy rain in southern Thailand killed 131 people.

A blizzard with 50-MPH (80-KPH) winds and temperatures below 0°F (-17.8°C) swept through the central United States in early January, killing 50 people.

A tornado destroyed a shopping mall in Mississippi, on January 10, killing 12 people and injuring about 200.

A tropical storm caused 30 deaths in January on land and off the coast of Mindanao, Philippines.

Prolonged drought in East Africa, starting in 1974 and lasting until June, affected about 800,000 people in the Ogaden region of Ethiopia and the adjacent area of Somalia, causing the death of an estimated 40,000 people.

The Nile flooded on February 20 and 21, inundating 1,000 acres (400 hectares) in Egypt and destroying 21 villages. At least 15 people died.

Monsoon rains caused flooding in northwestern India in July. About 400 square miles (1,040 square kilometers) were inundated, 14,000 homes were destroyed or badly damaged, and 300 people died.

Typhoon Phyllis struck the island of Shikoku, Japan, in August, killing 68 people. A week later it was followed by Typhoon Rita, which killed 26, injured 52, and left 3 unaccounted for.

Flood water inundated San'a, Yemen Arab Republic, in August, killing 80 people.

Monsoon rains caused floods up to 10 feet (3 meters) deep in the town of Bulandshahr, in Uttar Pradesh State, India, in September. At least 30 people drowned.

Hurricane Eloise, category 4, with winds up to 140 MPH (225 KPH) and heavy rain, struck Puerto Rico on September 16, causing widespread damage and killing 34 people. It then moved to Hispaniola, killing 25 people, Haiti and the Dominican Republic, then reached Florida, where 12 people died; finally Eloise moved to the northeastern United States, where damage was so severe that a state of emergency was declared.

Hurricane Olivia struck Mazatlán, Mexico, on October 24, killing 29 people.

1976

Hurricane-force winds of more than 100 MPH (160 KPH) (hurricane category 3) struck northern Europe on January 2 and 3, killing 26 people in England; 12 in Germany; and 17 in Denmark, Belgium, the Netherlands, Sweden, Austria, France, and Switzerland combined.

A tornado struck at least 12 villages in Faridpur District, Bangladesh, on April 10, killing 19 people and injuring more than 200.

Rains associated with Typhoon Olga caused widespread flooding in Luzon, Philippines, in May. At least 600,000 people were made homeless, and 215 were killed; damage was estimated at $150 million.

In July, floods covering millions of acres followed prolonged, heavy rain in Mexico, causing an estimated 120 deaths and leaving hundreds of thousands homeless.

The Ravi River, in northern Pakistan, overflowed on August 10, causing flooding that affected about 5,000 villages and killed more than 150 people.

Earthquakes caused tsunamis in the Philippines on August 17 in which more than 6,000 people died.

A tropical storm struck Hong Kong on August 25, killing at least 11 people, injuring 62, and leaving about 3,000 homeless.

Flood waters demolished an earth dam in Baluchistan Province, Pakistan, on September 5, causing floods that inundated more than 5,000 square miles (13,000 square kilometers) and swept away entire villages.

Typhoon Fran, category 3, struck southern Japan from September 8 to 13 with 100-MPH (160-KPH) winds and 60 inches (152.75 centimeters) of rain, killing 104 people, leaving 57 missing, and making an estimated 325,000 homeless.

Hurricane Liza, category 4, with 130-MPH (209-KPH) winds and 5.5 inches (13.97 centimeters) of rain, struck La Paz, Mexico, and on October 1 destroyed a 30-foot (9-meter) earth dam. A 5-foot (1.5-meter) wall of water swept through a shantytown, killing at least 630 people and leaving tens of thousands homeless.

Heavy rains caused flooding in Trapani, Sicily, on November 6, in which 10 people died.

In November, heavy rain caused widespread flooding in eastern Java, Indonesia, killing at least 136 people.

Heavy rain caused widespread flooding in Aceh, Sumatra, Indonesia, on December 20, killing at least 25 people.

1977

Heavy rain caused two rivers to overflow in southwestern Brazil in January. The floods caused 60 deaths and left about 3,500 people homeless.

States of emergency were declared in New York, New Jersey, and Ohio on January 28 due to blizzards and freezing. Several other states were declared disaster areas.

A cyclone with heavy rain struck Madagascar in February, destroying more than 230 square miles (596 square kilometers) of rice fields and 30,000 homes, and killing 31 people.

Blizzards throughout the northeastern United States on February 1 caused more than 100 deaths.

A blizzard in the central United States in March blocked 100 miles (160 kilometers) of interstate highway in South Dakota and killed nine people in Colorado, four in Nebraska, and two in Kansas.

A tornado struck Madaripur and Kishorganj, Bangladesh, on April 1, killing more than 600 people and injuring 1,500.

On April 4, tornadoes and floods affected West Virginia, Virginia, Alabama, Mississippi, Georgia, Tennessee, and Kentucky, killing 40 people and causing damage estimated to cost $275 million.

A cyclone, category 3, with 100-MPH (160-KPH) winds struck northern Bangladesh on April 24, killing 13 people and injuring nearly 100.

A tornado at Moundou, southeastern Chad, on May 20 killed 13 people and injured 100.

Floods following heavy rain in Khorramabad, Iran, on May 23 killed 13 people.

Floods in Iran on June 4, following heavy rain in the northwest of the country, killed at least 10 people.

A cyclone struck Oman in June, destroying 98 percent of the buildings on the island of Masirah, killing 2 people and injuring more than 40. Three days later, heavy rain caused flooding in Dhofar Province, where more than 100 people were killed and about 15,000 farm animals were carried away by the flood water.

Severe flooding followed heavy rain in southwestern France in July, killing 26 people and causing widespread damage to crops, livestock, and property.

Flooding and landslides caused by heavy rain in and around Seoul, South Korea, in July killed at least 200 people, injured more than 480, and left 80,000 homeless.

Overnight rain caused flooding in Johnstown, Pennsylvania, on July 20. About 70 people died, 30 were missing, and damage was estimated at $200 million.

Typhoon Thelma, category 4, with winds up to 120 MPH (193 KPH), struck Kaohsiung, Taiwan, on July 25, killing 31 people and destroying nearly 20,000 homes.

A typhoon struck Taipei, T'ao-yuan, and Nan-t'ou, Taiwan, on July 31, killing at least 38 people.

Following 12 inches (30 centimeters) of rain, flash floods struck Kansas City, Missouri, on September 13. At least 26 people died, and damage was estimated at more than $100 million.

In September, flooding followed 16 hours of continuous, heavy rain in Taipei, Taiwan, killing at least 14 people.

In October, widespread flooding in northwestern Italy caused 15 deaths and caused damage estimated at $350 million in Genoa and parts of Piedmont, Liguria, and Lombardy.

After 2.7 inches (6.9 centimeters) of rain fell in 15 hours on November 2 and 3, the Kifissos and Ilissos rivers, Greece, rose 6.5 feet (2 meters) and caused flooding in Athens and Piraeus, killing 26 people.

Prolonged, heavy rain caused the collapse of a dam near Toccoa, Georgia, on November 6, releasing a 30-foot (9-meter) wall of water that killed at least 39 people and injured 45.

Flash floods and landslides in Palghat, Kerala, India, on November 8 killed at least 24 people and injured 3.

Five days of heavy rain caused floods and landslides in northern Italy on November 10. Genoa and Venice were inundated. At least 15 people died, and thousands were left homeless.

A cyclone in Tamil Nadu, India, killed more than 400 people on November 12.

A typhoon struck the northern Philippines on November 14, killing at least 30 people and leaving nearly 50,000 homeless.

A cyclone and storm surge struck Andhra Pradesh, India, on November 19, washing away 21 villages and severely damaging 44. An estimated 20,000 people died, and more than two million were made homeless.

1978

Rainstorms, high tides, and 75-MPH (120-KPH) winds (hurricane category 1) caused extensive flooding and property damage along the east coast of the United Kingdom on January 12, killing 17 sailors when three ships sank, as well as 7 other people.

Floods in southeastern Colombia caused 20 deaths on January 13.

From January 13 to 17, floods in Rio de Janeiro, São Paulo, and Paraíba, in southeastern Brazil, killed 26 people and left thousands homeless.

A blizzard on January 25 and 26 brought winds gusting to 100 MPH (160 KPH), about 31 inches (78.7 centimeters) of snow, and temperatures down to -50°F (-45.5°C) in Ohio, Michigan, Wisconsin, Indiana, Illinois, and Kentucky. More than 100 people died, and damage was estimated at millions of dollars.

From January 26 to 30, severe flooding in East Java, Indonesia, caused 41 deaths.

From January 28 to 30, flooding caused 26 deaths in Transvaal, South Africa.

From February 5 to 7, a blizzard, with winds of 110 MPH (177 KPH) and 18-foot (5.5-meter) tides, brought 50 inches (127 centimeters) of snow to Rhode Island and eastern Massachusetts. At least 60 people died.

Severe flooding caused by heavy rain released a wall of water 20 feet (6 meters) high along a canyon in southern California on February 10. It destroyed the resort of Hidden Springs and caused the ports of Los Angeles and Long Beach to be closed. Zoo animals escaped when their cages were demolished, and 25 people were missing and feared dead.

A storm caused flooding and mudslides in southern California and northern Mexico on March 5. Five people were killed in California and 20 in Mexico, and 20,000 people were left homeless.

A tornado lasting about two minutes caused severe damage in northern Delhi, India, on March 17, killing 32 people and injuring 700.

The Zambezi River flooded in Mozambique on March 27, killing at least 45 people and leaving more than 200,000 homeless.

A tornado in Orissa State, India, on April 16 killed nearly 500 people and injured more than 1,000.

In April, a tornado in West Bengal, India, was believed to have killed 100 people.

Floods in Sri Lanka on May 15 killed 10 people and left thousands homeless.

Floods in West Atjeh, North Sumatra, Indonesia, on May 16 caused 21 deaths.

In mid-June, a week of torrential rain caused floods in South Korea in which 17 people died, 10 were missing, 2,000 were left homeless, and the damage was estimated at $400,000.

Heavy monsoon rains caused widespread summer flooding in India. Nearly 900 people died, hundreds of thousands were made homeless, the fall rice crop was almost completely destroyed, and damage was estimated at $100 million.

Floods along the border between Afghanistan and Pakistan on July 10 caused at least 120 deaths.

On July 26, monsoon floods in six Indian states caused nearly 190 deaths.

Floods in central Texas, due to rainfall at nearly 1 inch (2.54 centimeters) an hour, caused more than 25 deaths on August 1.

The Hab River, Pakistan, overflowed in August, causing floods that killed more than 100 people.

Flooding and mudslides in the northern Philippines in August killed 45 people and left thousands homeless.

Typhoon Rita, category 4, struck the Philippines on October 26. Nearly 200 people were killed, 60 were not accounted for, and about 10,000 homes were destroyed.

In November at least 144 people died when monsoons rains caused floods in Kerala and Tamil Nadu, southern India.

A cyclone struck Sri Lanka and southern India on November 23, killing at least 1,500 people. It destroyed more than 500,000 buildings and flooded about 45 villages.

1979

A blizzard on February 19 caused 13 deaths in New York and New Jersey.

Heavy rain caused floods and landslides in Flores Island, Indonesia, in March, killing 97 people, injuring 150, and leaving 8,000 homeless.

Cyclone Meli struck Fiji on March 27, killing at least 50 people and destroying about 1,000 homes.

A tornado with winds up to 225 MPH (360 KPH) moved through the Red River Valley on the border between Texas and Oklahoma on April 10. As it passed through Wichita Falls, Texas, it killed 59 people and injured 800.

A typhoon struck the Philippines on April 16 and 17, killing at least 12 people and causing damage estimated at $3.5 million.

A cyclone struck Andhra Pradesh and Tamil Nadu, India, on May 12 and 13, killing more than 350 people.

On June 13, floods following heavy rain at Balikpapan, Borneo, Indonesia, caused 13 deaths.

In June, floods up to 14 feet (4.2 meters) deep at Montego Bay, Jamaica, caused by heavy rain, killed at least 32 people and left 25 unaccounted for.

A flash flood at Valdepeñas, Spain killed at least 22 people on July 2.

A tsunami 6 feet (1.8 meters) high, caused by the collapse of the volcano Gulung Werung, struck Lomblem Island, Indonesia, on July 18, killing 539 people.

Heavy rain swelling the Machhu River, at Morvi, India, caused an earth dam to collapse on August 11, sending a wall of water 20 feet (6 meters) high through the town and killing up to 5,000 people.

Tornadoes forming in the midwestern and New England states moved eastward, crossing the Atlantic and reaching the Irish Sea in August, during the Fastnet Race between England and Ireland. Of 306 yachts entering the race, only 85 finished; 23 vessels sank or were abandoned, and 18 people died.

Typhoon Judy brought intense rain that caused flooding in southern South Korea on August 25 and 26, in which nearly 60 people died and about 20,000 were made homeless.

Hurricane David, category 5, with winds up to 150 MPH (240 KPH), struck the Caribbean and east coast of the United States in late August and early September, affecting the Dominican Republic, Dominica, Puerto Rico, Haiti, Cuba, and the Bahamas, as well as Florida, Georgia, and New York. Altogether more than 1,000 people were killed, and the damage was estimated at billions of dollars.

Hurricane Frederic struck 100 miles (160 kilometers) of the U.S. coast in Florida, Alabama, and Mississippi in early September. About eight people were killed. The swift evacuation of nearly 500,000 people saved many lives.

The Brahmaputra River overflowed on October 13, causing widespread flooding in Assam, India. At least 13 people drowned.

Two tsunamis up to 10 feet (3 meters) high, caused by a submarine landslide, struck a 60-mile (96-kilometer) stretch of the French Mediterranean coast on October 16. Eleven people were swept away in Nice and one in Antibes, and were presumed dead.

A severe storm caused flash floods on October 17 near Groblersdal, South Africa. The Elands River overflowed and a large dam broke, carrying away at least 20 people.

Typhoon Tip, with winds up to 55 MPH (88 KPH), caused widespread damage throughout Japan on October 19. At least 36 people died.

A blizzard with 70-MPH (112-KPH) winds and heavy falls of snow affected Colorado, Nebraska, and Wyoming on November 21, killing at least 10 people.

The Playonero River overflowed on November 25, sending flood water and mud through the towns of El Playón and Lebrija, Colombia, killing 62 people.

1980

Cyclone Hyacinthe struck the island of Réunion in January, killing at least 20 people.

Floods and huge mudslides followed prolonged, heavy rain in California, Arizona, and Mexico in February, causing 36 deaths and damage estimated at $500 million.

A blizzard on March 2 caused at least 36 deaths in North Carolina, South Carolina, Ohio, Missouri, Tennessee, Pennsylvania, Kentucky, Virginia, Maryland, and Florida.

Heavy rain caused floods and landslides in several parts of Turkey on March 27 and 28, killing at least 75 people.

Heavy rain caused floods, mudslides, and rock slides in central Peru in April, leaving at least 90 people missing and feared dead.

Cyclone Wally struck Fiji in April, causing floods and landslides that killed at least 13 people and left thousands homeless.

Tornadoes passed through Kalamazoo, Michigan, on May 13, killing 5 people and injuring at least 65.

In July, floods in Gujarat State, India, caused 37 dams to overflow; 11 people died.

Typhoon Joe struck North Vietnam on July 23, killing more than 130 people and making about three million homeless.

In India, July and August monsoon floods inundated 7,500 square miles (19,425 square kilometers), killed at least 600 people, and caused damage estimated at more than $131 million.

Hurricane Allen, category 5, with winds of 175 MPH (280 KPH) gusting to 195 MPH (314 KPH), struck Barbados, St. Lucia, Haiti, the Dominican Republic, Jamaica, Cuba, and the southeast United States in August. More than 270 people died, most of them in Haiti.

Two Chinese cities were flooded in August when the Dongting Hu (Lake), fed by the Yangtze River, overflowed. Thousands of people died.

In August and September, monsoon rains caused flooding and landslides in West Bengal, India, killing nearly 1,500 people.

Heavy rain caused a dam to burst above Arandas, Mexico, causing floods from September 1 to 3 in which at least 100 people were dead or left missing.

Typhoon Orchid struck South Korea on September 11, killing 7 people and causing more than 100 fishermen to be lost at sea.

Typhoon Ruth struck Vietnam on September 15 and 16, killing at least 164 people.

Heavy rain caused flash floods in Orissa, India, in September. The floods burst a dam, flooding two towns to a depth of 10 feet (3 meters). About 200 people drowned, and at least 300,000 were marooned.

The Guaire River, near Caracas, Venezuela, overflowed in September, causing floods that brought the city to a standstill and killed at least 20 people.

In September, floods in northwestern Bangladesh killed 655 people.

A cyclone struck Maharashtra State, India, in September, killing at least 12 people and injuring 25.

In October, widespread monsoon floods in Thailand killed 28 people.

1981

The Buffels River, South Africa, overflowed on January 24 and 25, sending a wall of water 6 feet (1.8 meters) high through the town of Laingsburg and killing at least 200 people.

Drought in northeastern Brazil ended in March and early April with 10 days of continuous rain that caused floods in which 30 people drowned and about 50,000 were left homeless.

In April, week-long floods in Colombia killed 65 people and left 14,000 homeless.

A tornado at Noakhali, Bangladesh, on April 12 killed about 70 people, injured 1,500, and destroyed 15,000 homes.

A tornado passed through four villages in Orissa State, India, on April 17, killing more than 120 people, injuring hundreds, and destroying 2,000 homes.

In April, prolonged, heavy rain caused floods and mudslides in Venezuela; at least 27 people died.

On May 3, floods in Khorasan Province, Iran, killed or injured 100 people.

In May, floods and mudslides in Java, Indonesia, killed 127 people, left 170 unaccounted for, and injured 38.

Shoal Creek, near Austin, Texas, overflowed because of flash floods on May 25, sending flood water through the city, killing 10 people and leaving 8 missing.

Typhoon Kelly struck the central Philippines on July 1, causing floods and landslides and killing about 140 people.

Monsoon rains caused the Yangtze River and its tributaries to flood from July 12 to 14. About 1,300 people were killed or left missing, more than 28,000 were injured, and 1.5 million were left homeless. Damage was estimated at $1.1 billion.

Flash floods in northern Nepal on July 13 killed nearly 100 people.

Typhoon Maury struck northern Taiwan on July 19, causing floods and landslides and killing 26 people.

On July 19, floods in Assam, Uttar Pradesh, and Rajasthan, India, killed about 500 people and left 100,000 homeless.

The Salamina River, Colombia, overflowed on August 17, flooding the town of Saravena and leaving 150 people dead or missing.

In August, floods in Sichuan Province, China, killed 15 people.

Typhoon Tad, category 1, with winds up to 80 MPH (128 KPH), struck central and northern Japan on August 23, killing 40 people and leaving 20,000 homeless.

Typhoon Agnes struck South Korea on September 1, bringing 28 inches (71.12 centimeters) of rain in two days and leaving 120 people dead or missing.

Floods in El Eulma, Algeria, on September 3 killed 43 people.

Winds up to 80 MPH (128 KPH) (hurricane category 1) struck the United Kingdom on September 21, killing at least 12 people.

Typhoon Clara struck Fujian Province, China, on September 21, destroying 130 square miles (337 square kilometers) of rice crops.

Flash floods in Nepal on September 29 left about 500 people dead or missing.

In October, floods and landslides in Sichuan Province, China, killed 240 people.

Storms caused floods in northern Mexico on October 7, in which 65 people died.

A blizzard in Michigan and Minnesota on November 19 and 20 caused at least 17 deaths.

Typhoon Irma, category 5, with winds of 140 MPH (225 KPH) struck the Philippines on November 24, causing great destruction in the coastal towns of Garchitorena and Caramoan. More than 270 people died, and 250,000 were left homeless. Damage was estimated at $10 million.

In December, monsoon rains caused flash floods in Thailand; at least 37 people died.

Floods and mudslides in Brazil on December 3 killed more than 40 people.

A typhoon struck Bangladesh and India on December 11, killing at least 27 people.

1982

Five cyclones called Benedict, Frida, Electra, Gabriel, and Justine struck Madagascar between January and March. More than 100 people were killed and 117,000 left homeless.

Floods on January 4 killed 90 people in Nariño Province, Colombia.

Floods and landslides on January 5 caused 15 deaths in Rio de Janeiro, Brazil.

Blizzards in western Europe from January 9 to 12 caused at least 23 deaths. Wales was completely cut off from England by snowdrifts 12 feet (3.6 meters) deep.

The Chuntayaco River, in western Peru, burst its banks along a 60-mile (96-kilometer) stretch on January 23 and 24, carrying away 17 villages. At least 600 people died, and 2,000 were left missing.

In March, floods in Santa Cruz, Bolivia, caused widespread damage to crops. About 50 families were presumed drowned.

Typhoons Mamie and Nelson struck the Philippines in March, killing at least 90 people and leaving 17,000 homeless.

In April, heavy rain caused floods and landslides in Cuzco Province, Peru. About 200 people were believed killed.

Tornadoes moved through Ohio, Texas, Arkansas, Mississippi, and Missouri on April 2 and 3, killing 31 people—10 of them in Paris, Texas, and 14 in Arkansas.

A blizzard on April 6 killed 33 people in the northern United States.

A cyclone, category 4, with 124-MPH (199-KPH) winds, struck Burma on May 4, killing 11 people and leaving 7,200 families homeless.

Tornadoes moved through Kansas, Oklahoma, and Texas on May 11 and 12, killing at least seven people. Damage at Altus Air Force Base, in Oklahoma, was estimated at $200 million.

In May, severe flooding in Guangdong Province, China, killed at least 430 people and destroyed about 46,000 homes.

Severe flooding in May killed 75 people in Nicaragua and 125 in Honduras and caused $200 million of damage.

Flooding in Hong Kong on May 29 killed at least 20 people.

A tornado at Marion, Illinois, killed at least 10 people on May 29.

Monsoon rains caused flooding in Sumatra, Indonesia, on June 3, killing at least 225 people and leaving about 3,000 homeless.

Winds of up of 137 MPH (220 KPH) (hurricane category 5) struck Orissa, India, on June 4, killing 200 people and leaving about 200,000 homeless.

Flooding in Connecticut on June 5 and 6 killed at least 12 people.

On June 26 and 27, winds up to 90 MPH (144 KPH) (hurricane category 2) in Paraña and São Paulo, Brazil, killed at least 43 people and injured 500.

In June, floods in Fujian Province, China, killed 75 people.

In July, monsoon rains caused floods and landslides in southern Japan, killing 245 people and leaving 117 unaccounted for.

A typhoon struck the coast of South Korea on August 12 and 13, causing flash floods and landslides and leaving 38 people dead, 26 missing, 100 injured, and 6,000 homeless.

Typhoon Cecil struck South Korea in August, killing at least 35 people, leaving 28 unaccounted for, and causing more than $30 million in damage.

In September, monsoon flooding in Orissa, India, caused at least 1,000 deaths and left five million people marooned on high ground and roofs, dependent on supplies dropped by air. Eight million people were displaced, and more than 2,000 cattle killed.

Typhoon Judy, category 3, with winds of 110 MPH (177 KPH), struck Japan on September 11 and 12, killing 26 people, injuring 94, leaving 8 missing, and causing extensive damage.

Floods and mudslides in El Salvador from September 17 to 21 killed at least 700 people, injured 18,000, and left 55,000 homeless. In Guatemala, the storms that caused the floods killed 615.

Hurricane Paul, category 4, with winds of 120 MPH (193 KPH), struck Sinaloa, Mexico, on September 30, leaving 50,000 people homeless.

A hurricane struck Nghe Tinh Province, Vietnam, in October, killing hundreds of people and leaving nearly 200,000 homeless.

Drought in Indonesia, lasting four months, led to outbreaks of cholera and dengue fever in October; more than 150 people died.

A category 4 typhoon, with winds of 120 MPH (193 KPH), struck the provinces of Isabela, Kalinga-Apayao, and Cagayan, Philippines, on October 14 and 15, killing 68 people and leaving tens of thousands homeless.

A category 4 hurricane, with winds of 125 MPH (200 KPH), struck Gujarat, India, on November 8, killing at least 275 people and destroying 30,000 homes.

December flooding in Arkansas, Illinois, and Missouri killed 20 people, left 4 missing, and caused at least $500 million of damage.

Blizzards, storms, and tornadoes struck the western United States in December, killing 34 people.

1983

January flooding in Ecuador caused 30 deaths and caused about $90 million of damage.

The Arrucadas River, Brazil, overflowed on January 3, flooding the city of Belo Horizonte and killing 40 people.

A blizzard on February 11 and 12 delivered at least 2 feet (60 centimeters) of snow to every city in the northeastern United States and caused at least 11 deaths.

Blizzards near Alayh, Lebanon, from February 18 to 22 killed 47 people, many of them frozen while trapped in their cars.

The Pirai River, in Santa Cruz, Bolivia, overflowed in March, leaving 96 people dead or missing.

In March, flooding in Guangdong Province, China, killed at least 27 people.

Floods and mudslides in Peru and Bolivia on March 20 and 21 killed at least 260 people and left hundreds missing.

A cyclone struck West Bengal, India, in April, killing 76 people, injuring about 1,500, and leaving 6,000 homeless.

A tornado in Fujian Province, China, on April 11 killed 54 people.

A cyclone on April 12 caused damage in about 21 coastal villages near Calcutta, India, killing at least 50 people, injuring 1,500, and leaving 6,000 homeless.

On April 14, floods and landslides in Piura and Tumbes, Peru, killed 37 people.

A tornado struck Khulna, Bangladesh, on April 26, killing 12 people and injuring 200.

A flash flood near Chepén, Peru, on April 30 swept over a road bridge, knocking several vehicles into the Chaman River. At least 50 people drowned.

Flooding occurred in May when major rivers in northern France and southern Germany overflowed. Altogether 25 people died.

May flooding in the basins of the Paraguay and Paraná rivers in Argentina, Brazil, and Paraguay killed 23 people and caused damage estimated at $338 million.

In May, tornadoes in central Vietnam killed more than 76 people.

At least 59 tornadoes, with storms that produced flash floods, moved through Texas, Tennessee, Missouri, Georgia, Louisiana, Mississippi, and Kentucky from May 18 to 20. At least 24 people died and 350 homes were destroyed in Houston, Texas.

A tsunami caused by a Richter magnitude 7.7 earthquake struck northern Honshu, Japan, on May 26, killing at least 58 people.

In June, widespread flooding and landslides in Ecuador killed at least 300 people.

Floods and landslides in Taiwan on June 5 killed 24 people.

June flooding in Gujarat, India, left at least 935 people dead or missing.

The Yangtze River, China, overflowed in late June and early July, causing extensive flooding that was believed to have caused hundreds of deaths.

On July 23, floods and landslides in Masuda, Japan, killed 82 people.

In July, flooding in Banggai, Indonesia, killed 11 people and left 2,000 homeless.

In August, flooding in Bangladesh killed 41 people.

Hurricane Alicia, category 4, with winds of 115 MPH (185 KPH), struck southern Texas on August 18, causing extensive damage in Galveston and Houston. At least 17 people died, and damage was estimated at $1.6 billion.

On August 26, flooding in the Basque region on the border between France and Spain left 33 people dead and 13 missing.

In September, monsoon rains in India killed more than 400 people.

In September, flooding in Papua New Guinea killed 11 people and caused $11.9 million of damage.

Typhoon Forest struck the Japanese islands of Honshu, Kyushu, and Shikoku on September 29, bringing up to 19 inches (48.3 centimeters) of rain and leaving 16 people dead, 22 missing, and 30,000 homes flooded.

In September, flooding in Bangladesh killed 61 people, 6 of whom died from snake bites while climbing trees away from the water.

In late September and early October, flooding in southern Arizona left a number of towns under several feet of water and caused 13 deaths.

Flooding in Uttar Pradesh, India, on October 14 killed 42 people.

A cyclone struck Bangladesh on October 15, destroying 1,000 homes in Chittagong and killing at least 25 people.

In October, widespread flooding in Thailand caused 18 deaths and extensive damage in Bangkok.

Hurricane Tico struck the coast of Mazatlán, Mexico, on October 20, killing 105 fishermen whose boats were lost at sea.

A blizzard on November 28 in Wyoming, Colorado, South Dakota, Nebraska, Kansas, Minnesota, and Iowa killed at least 56 people.

In December, monsoon rains in Malaysia caused at least 10 deaths from drowning and necessitated the evacuation of 15,000 people.

1984

Gales and snowstorms crossed northern Europe from January 12 to 16, causing at least 22 deaths in the United Kingdom.

A hurricane struck Swaziland on January 30 and 31, killing at least 13 people.

Cyclone Domoina struck Mozambique, South Africa, and Swaziland from January 31 to February 2, causing severe flooding that killed at least 124 people and left thousands homeless.

Blizzards in the United States on February 4 killed at least 33 people.

Blizzards in western Europe on February 7 killed 13 people.

In February, monsoon rains in Java, Indonesia, killed 26 people.

Blizzards from Missouri to New York on February 28 caused at least 29 deaths.

Blizzards in the eastern United States on March 9 killed 23 people.

Blizzards in New England on March 14 killed at least 11 people.

From March 19 to 23, blizzards and thunderstorms in the western United States killed 27 people.

Tornadoes in North and South Carolina killed more than 70 people on March 28.

A cyclone, category 5, with winds of 150 MPH (240 KPH), struck Mahajanga, Madagascar, on April 12, destroying about 80 percent of the town and killing at least 15 people.

Tornadoes in Water Valley, Mississippi, killed 15 people on April 21.

On April 26, a tornado in Oklahoma killed 11 people and destroyed more than half the buildings in Morris, and killed 3 people elsewhere in the state.

From May 6 to 9, tornadoes and thunderstorms, with serious flooding in Appalachia and Kentucky, Louisiana, Tennessee, Ohio, Maryland, and West Virginia, killed at least 14 people and left 6,000 homeless.

Floods and landslides in Bangladesh and India from May 13 to 16 killed at least 136 people.

Floods caused by overnight rain in Tulsa, Oklahoma, on May 27 killed at least 12 people and left thousands homeless.

In May, flooding in Rio Grande do Sul, Brazil, killed 17 people and left nearly 10,000 homeless.

In late May and early June, floods in the northeastern United States caused 18 deaths and left thousands of people homeless.

In June, flooding in northeastern India killed at least 38 people.

Monsoon floods in Bangladesh and India killed an estimated 200 people in June.

Flooding and landslides in Taipei, Taiwan, on June 3 caused 25 deaths.

Storms with 49 tornadoes caused extensive damage in the Midwest on June 8. Barneveld, Wisconsin, was totally demolished, and 9 people died.

Tornadoes in Russia on June 9 and 10 caused widespread damage and hundreds of deaths in the towns of Ivanovo, Gorky, Kalinin, Kostroma, and Yaroslavl.

Flooding in South Korea from July 4 to 7 killed at least 14 people and left nearly 2,000 homeless.

In July, floods and landslides around Recife, Brazil, killed at least 13 people and left 1,000 homeless.

Floods in Seoul, South Korea, from August 31 to September 3 left 81 people dead, 36 missing, and caused more than $7 million of damage.

Typhoon Ike struck the Philippines on September 2 and 3, killing more than 1,300 people and leaving 1.12 million homeless.

Typhoon Ike struck the coast of Guangxi Zhuang, China, on September 6, causing widespread damage and killing 13 people whose fishing boats were lost at sea.

Floods and landslides in Nepal killed more than 150 people in September.

In October, floods in central Vietnam killed 33 people and left more than 38,000 families homeless.

A tornado at Maravilha, Brazil, on October 9 killed at least 10 people.

Typhoon Agnes, category 5, with winds of 185 MPH (297 KPH), struck the central Philippines in November, killing at least 300 people, leaving 100,000 homeless, and causing $40 million of damage.

Flooding in Colombia killed at least 40 people in November.

Hurricane-force winds caused widespread damage on November 24 in England, Germany, the Netherlands, France, and Belgium; at least 14 people died.

1985

In January, flooding in Brazil killed at least 71 people and left thousands homeless.

Flooding in Algeria on January 5 killed at least 26 people.

Cyclones Eric and Nigel struck Viti Levu, Fiji, on January 22, killing 23 people.

Monsoon rains in Indonesia on February 21 caused a landslide in Lombok, killing at least 11 people, and floods in Java that killed 10.

Floods in northeastern Brazil killed 27 people in April.

A cyclone with a 10- to 15-foot (3- to 4.5-meter) storm surge struck islands at the mouth of the Meghna (Ganges) River, Bangladesh, on May 25. The death toll was set at 2,540, but might have been as high as 11,000.

Floods inundated Buenos Aires, Argentina, on May 30 and 31, killing at least 14 people and forcing 90,000 to leave their homes.

Tornadoes crossed Pennsylvania, Ohio, and New York, and Ontario, Canada, on May 31, killing at least 88 people and almost destroying several towns in Pennsylvania.

Floods in Guangxi Province, China, killed 64 people on June 6.

In June, monsoon rains caused floods and landslides in western India, leaving at least 46 people dead and 25,000 homeless.

In June, monsoon rains caused flooding in the northern Philippines, leaving at least 65 people dead and more than 100,000 homeless.

Typhoon Irma struck Japan on July 1, causing 19 deaths and extensive damage in Numazu and Tokyo.

Monsoon floods in Punjab, India, killed 87 people in July.

A typhoon struck Zhejiang Province, China, on July 30, killing 177 people and injuring at least 1,400.

Dikes failed at the Yalu River, near the border between North Korea and China; the Yalu overflowed in late July and early August, causing floods that destroyed two villages near Dandong, China, and killing 64 people.

Typhoons and heavy rain in China killed more than 500 people in August and left 14,000 homeless.

Typhoon Pat, category 4, with winds of up to 124 MPH (200 KPH), struck Kyushu, Japan, on August 30, leaving 15 people dead and 11 missing.

Two typhoons struck Thailand in October, causing floods in which 16 people died.

Typhoons struck Bangladesh on October 16 and 17, killing 12 people.

Floods in India on October 18 and 19 caused 78 deaths.

Typhoon Dot struck Luzon, Philippines, on October 19, destroying 90 percent of the buildings in the city of Cabanatuan, killing at least 63 people, and causing $5.3 million of damage.

In October, a tropical storm caused floods in Louisiana, leaving seven people dead and eight missing, and causing $1 billion of damage.

Floods in West Virginia, Virginia, Maryland, and Pennsylvania, following 20 inches (50.8 centimeters) of rain in 12 hours, killed at least 49 people.

A blizzard in the northwestern United States killed at least 33 people in November.

Hurricane Kate struck Cuba and Florida from November 19 to 21, killing at least 24 people.

Blizzards in Michigan, South Dakota, Iowa, Minnesota, and Wisconsin killed 19 people in December.

In December, floods in Saudi Arabia left 32 people dead and 31 missing.

1986

Floods and landslides in Sri Lanka killed 43 people in January.

In January, monsoon rains caused floods and landslides in Indonesia; 19 people died.

Lake Titicaca, Peru, overflowed its banks in March, flooding the city of Puno and leaving at least 12 people dead and 28 missing.

Cyclone Honorinnia struck Madagascar on March 17, destroying 80 percent of the buildings in Toamasina and leaving 32 people dead and 20,000 homeless.

Hurricane-force winds struck western Europe on March 24, leaving at least 17 people dead and 19 missing.

A cyclone struck Rajasthan, India, on May 16, killing 11 people.

Typhoon Namu struck the Solomon Islands on May 19, killing more than 100 people and leaving more than 90,000 homeless.

In June, floods in central Chile killed 10 people and left 35,000 homeless.

Typhoon Peggy struck the northern Philippines on July 9, causing floods, mudslides, and extensive damage. More than 70 people were killed.

Typhoon Peggy caused widespread flooding in southeastern China on July 11. More than 170 people died, at least 1,250 were injured, and more than 250,000 homes were destroyed.

A typhoon struck Taiwan on August 22, leaving 22 people dead, 9 missing, and more than 110 injured.

Hurricane Charley struck the British Isles on August 25, causing at least 11 deaths.

In August, monsoon rains caused flooding in Andhra Pradesh, India, in which more than 200 people died.

A typhoon struck Vietnam on September 4, killing 400 people and injuring 2,500.

Typhoon Abby struck Taiwan on September 19, killing 13 people and causing $80 million of damage.

Floods in and around Manila, Philippines, on October 6 killed 14 people and forced nearly 60,000 to leave their homes.

1987

A blizzard from Maine to Florida on January 22 caused at least 37 deaths.

Floods inundated São Paulo, Brazil, in January, causing at least 75 deaths.

In February, floods and mudslides in Peru destroyed part of the town of Villa Rica and killed more than 100 people.

February floods in the Republic of Georgia left 30 people dead and 6 missing.

Cyclone Uma struck Vanuatu on February 7, killing at least 45 people.

The area around Lima, Peru, was flooded on March 9 when dams weakened by rain collapsed. More than 100 people died, and 25,000 were left homeless.

A tornado at Saragosa, Texas, on May 22 killed 29 people.

From July 12 to 16, floods in central Chile caused 16 deaths, 12 of them when a bridge north of Santiago was destroyed.

A flash flood at Le Grand-Bornand, France, on July 14—caused when the Borne River broke its banks—washed away a vacation campsite, killing at least 30 people.

Typhoon Thelma struck South Korea on July 15, causing floods, landslides, and mudslides. At least 111 people died, and 257 were left missing.

Floods and landslides in northern Italy on July 18 killed at least 14 people, 12 of them at a mountain hotel that was destroyed.

Floods and landslides in Chunchon Province, South Korea, on July 21 and 22 left 100 people missing, feared dead.

The Boojhan River, Iran, overflowed on July 24, causing floods in which at least 100 people died.

Seoul, South Korea, was flooded on July 27, and at least 74 people died.

Typhoon Alex struck Zhejiang Province, China, on July 28, triggering a huge landslide. There was widespread damage, and at least 38 people died.

A tornado in Heilongjiang Province, China, on July 31 caused extensive damage in 14 towns and left 16 people dead, 13 missing, and more than 400 injured.

Five tornadoes, with 60-MPH (96-KPH) winds, struck a trailer park and nearby industrial center at Edmonton, Alberta, Canada, on July 31 and killed more than 25 people.

Floods in Bangladesh killed more than 1,000 people in August.

Floods near Maracay, Venezuela, killed about 500 people in September.

Floods in Natal Province, South Africa, from September 25 to 29 left 174 people dead, 86 missing, more than 50,000 homeless, and caused $500 million of damage.

Heavy rain caused rock and mud to slide down on Villa Tina, Medellín, Colombia, on September 27, leaving at least 175 people dead and 325 missing, believed buried beneath rubble.

Floods in northern Bihar, West Bengal, Uttar Pradesh, and Assam, India, in September killed more than 1,200 people.

Hurricane-force winds struck England on October 15, killing 13 people and causing $1 billion of damage.

Typhoon Lynn struck Taiwan on October 24, destroying 200 homes.

Cyclone-force winds in Andhra Pradesh, India, from November 3 to 6 killed at least 34 people.

Typhoon Nina, with a storm surge, struck the Philippines on November 26, killing 500 people in Sorsogon Province, Luzon.

Blizzards in the midwestern United States from December 12 to 16, generating tornadoes in Arkansas, caused 73 deaths.

Floods and landslides in Sulawesi, Indonesia, killed at least 92 people on December 25.

Floods in Minas Gerais, Brazil, killed at least 12 people in December.

1988

Blizzards in the midwestern and eastern United States killed at least 33 people from January 2 to 8.

Mud and rocks, flowing after heavy rain, killed at least 30 people in Huanoco Province, Peru, on February 3.

In February, floods, landslides, and mudslides in Rio de Janeiro State, Brazil, killed more than 280 people and injured 600.

Flash floods in the Orange Free State, South Africa, killed at least 12 people on February 22.

Floods in Kenya on April 23 and 24 killed at least 13 people.

On May 22, flash floods in Fujian Province, China, killed 72 people and injured 200.

In May, floods in southeastern China caused at least 149 deaths.

June floods in Cuba killed at least 21 people.

Flash floods at Ankara, Turkey, killed 13 people on June 12.

On July 29 and 30, flash floods in Zhejiang Province, China, left 264 people dead and 50 missing, feared drowned.

August floods in Sudan caused when the Nile burst its banks, killed at least 90 people and left two million homeless.

Monsoon rains left 75 percent of Bangladesh inundated by floods in late August and September. More than 2,000 people died, and many more suffered waterborne diseases. At least 30 million people were left homeless.

September floods in southern China killed at least 170 people and left 110,000 homeless.

Hurricane Gilbert, category 5, struck the Caribbean and Gulf of Mexico from September 12 to 17. It caused extensive damage in

Jamaica and then moved toward the Yucatán Peninsula, killing about 200 people in Monterrey, Mexico, and causing $10 billion in damage. The hurricane killed at least 260 people and generated nearly 40 tornadoes in Texas.

Flash floods in the village of Darbang, western Nepal, on September 22 killed at least 87 people.

In September, flash floods in southern Ethiopia killed at least 81 people and left 2,240 homeless.

Widespread floods in northwestern India in late September and early October killed an estimated 1,000 people.

Floods in northern Vietnam from October 10 to 18 killed at least 27 people.

Hurricane Joan struck the Caribbean coast from October 22 to 27, causing severe damage in Nicaragua, Costa Rica, Panama, Colombia, and Venezuela and killing at least 111 people. The storm weakened and became Tropical Storm Miriam, which then struck El Salvador, where it left 3,000 people homeless.

Typhoon Ruby struck the Philippines on October 24 and 25, causing flooding and mudslides, killing about 500 people, and causing $52 million of damage.

Typhoon Skip struck the Philippines on November 7, killing at least 129 people.

In November and December, monsoon rains caused floods in southern Thailand; more than 400 people died.

A cyclone struck Bangladesh and eastern India on November 29, killing up to 3,000 people.

Floods and landslides in Java, Indonesia, killed at least 40 people in December.

1989

Cyclone Firinga, category 4, with winds of more than 125 MPH (200 KPH), struck Réunion on January 28 and 29, killing at least 10 people and leaving 6,000 homeless.

Floods in central Peru killed 57 people in February.

Hurricane-force winds in Spain on February 25 and 26 killed at least 12 people.

March floods in Yemen killed at least 23 people.

Floods, landslides, and avalanches caused by heavy rain killed more than 50 people in the Republic of Georgia on April 19.

On April 26, a tornado in Bangladesh struck more than 20 villages and left up to 1,000 people dead, 12,000 injured, and nearly 30,000 homeless.

On May 6, tornadoes and floods killed 23 people and injured more than 100 in Texas, Virginia, North Carolina, Louisiana, South Carolina, and Oklahoma.

Typhoon Cecil struck Vietnam on May 25 and 26, destroying 36,000 homes and leaving 140 people dead and 600 missing.

A cyclone struck Bangladesh and eastern India on May 27, killing 200 people.

Typhoon Brenda struck southern China in May, killing 26 people.

In June, monsoon rains caused floods in Sri Lanka, killing 300 people and leaving 125,000 homeless.

Floods in Sichuan Province, China, killed more than 1,300 people in June and July.

In June, floods and landslides in Ecuador killed 35 people and left about 30,000 homeless.

Blizzards in western China killed at least 67 people in June and July.

Typhoon Gordon struck Luzon, Philippines, on July 16, killing 33 people.

Typhoon Irving struck Thanh Hoa Province, Vietnam, on July 24, killing at least 200 people.

Typhoon Judy struck South Korea in July, killing at least 17 people.

Monsoon rains in July caused floods in which 81 people died in South Korea; 750 in India, with 2,000 left missing; 17 in Pakistan; 200 in Bangladesh; and 1,500 in China.

Typhoon Sarah struck Taiwan on September 11, breaking a Panamanian ship in half and killing 13 people.

Typhoon Vera struck Zhejiang Province, China, on September 16, leaving 162 people dead, 354 missing, and 692 injured.

Hurricane Hugo, category 5, with winds up to 140 MPH (224 KPH) and gusts up to 220 MPH (355 KPH), struck the Caribbean and east coast of the United States on September 17 to 21. Hugo reached Guadeloupe, killing 11 people, and Dominica in the Leeward Islands (Lesser Antilles) on September 17, then St. Croix, St. John, St. Thomas, and smaller islands in the U.S. Virgin Islands, and Puerto Rico on September 19. On Montserrat, 10 people died; 6 died in the Virgin Islands and 12 in Puerto Rico. Hugo then turned north and weakened to category 4, striking Charleston, South Carolina, on September 21, killing one person when a house collapsed, and Charlotte, North Carolina, on September 22, where one child died. Hugo then crossed the Blue Ridge Mountains and in the afternoon crossed Virginia, killing two people, and Pennsylvania. Winds of 81 MPH (130 KPH) were recorded in Virginia. Awendaw, South Carolina, was hit by a tidal surge, and tornadoes forming part of the storm caused damage on several islands and in North Carolina. Almost everyone in Montserrat was made homeless, and in Antigua 99 percent of homes were destroyed. In St. Croix, 90 percent of the population was left homeless, and 80 percent in Puerto Rico and Folly Beach, South Carolina. The hurricane caused damage costing $10.5 billion in the United States.

Typhoon Angela struck the Philippines in October, killing at least 50 people.

Three typhoons struck Hainan Province, China, from October 2 to 13, killing 63 people and injuring more than 700.

Typhoon Dan struck the Philippines on October 10, killing 43 people and leaving 80,000 homeless.

Typhoon Elsie struck the Philippines on October 19, killing 30 people and leaving 332,000 homeless.

Typhoon Gay struck Thailand on November 4 and 5, killing 365 people and damaging or destroying 30,000 homes.

A cyclone struck southern India on November 9, killing 50 people.

A tornado in Huntsville, Alabama, on November 15 killed 18 people and destroyed 119 homes.

December floods in Brazil left 35 people dead and 200,000 homeless.

1990

A cyclone struck Madagascar in January, killing at least 12 people. Floods in Tunisia from January 20 to 24 killed 30 people and left more than 9,500 homeless.

Hurricane-force winds struck Europe on January 25, killing 45 people in the United Kingdom, 19 in the Netherlands, 10 in Belgium, 8 in France, 7 in Germany, and 4 in Denmark.

Floods and landslides in Java, Indonesia, killed more than 130 people on January 27 and 28.

Hurricane-force winds struck France and Germany on February 3, killing 29 people and tearing tiles from the roof of Chartres Cathedral.

Hurricane-force winds struck Europe on February 26, killing at least 51 people.

Flooding in Kenya and Tanzania in March and April killed 140 people and left 25,000 Tanzanians homeless.

Flooding in Rio de Janeiro, Brazil, killed 11 people on April 18.

May flooding in Russia, caused when melting snow made the Belaya River overflow, inundated 130 villages and killed 11 people.

Floods in Texas, Oklahoma, Louisiana, and Arkansas killed 13 people in May.

A cyclone struck Andhra Pradesh, India, on May 9, killing at least 962 people and leaving thousands unaccounted for.

Tornadoes in Indiana, Illinois, and Wisconsin killed 13 people on June 2 and 3.

On June 14, a flash flood in Shadyside, Ohio, demolished homes and killed at least 26 people.

Floods in Hunan and Jiangxi Provinces, China, in June, destroyed 16,000 homes and killed more than 100 people.

Floods in Turkey killed 48 people on June 20.

Typhoon Ofelia struck the Philippines, Taiwan, and China on June 23 and 24, killing a total of 57 people.

Floods in Yunnan Province, China, killed 108 people in July.

Floods in Bangladesh killed 26 people in July.

In July, floods around Lake Baikal, Siberia, caused extensive damage and an unknown number of deaths.

Floods and landslides in Kyushu, Japan, killed 24 people on July 2.

In August, a typhoon struck Guangdong and Fujian provinces, China, killing 108 people.

A hurricane struck Mexico in August, causing floods in which 23 people drowned.

Typhoon Yancy struck the Philippines and China in August, killing 216 people in Fujian and Zhejiang provinces, China, and 12 in the Philippines.

Floods in the Chitwan National Park, Nepal, drowned 20 people in August.

On August 28, a tornado at Plainfield, Illinois, killed 29 people and injured 300.

Typhoon Abe struck Zhejiang Province, China, on August 31, killing 48 people.

On September 11 and 12, floods and landslides in Seoul, South Korea, left 83 people dead and 52 missing.

Typhoon Flo struck Honshu, Japan, on September 16 and 17, killing 32 people.

On September 22 and 23, flash floods in Chihuahua, Mexico, left 45 people dead, 30 missing, and 5,000 homeless.

Floods in Bangladesh killed 14 children in September.

In October, storm surges in the Bay of Bengal left 50 people dead and at least 3,000 fishermen missing in Bangladesh.

A typhoon struck Vietnam on October 23, killing 15 people and leaving thousands homeless.

Typhoon Mike struck the Philippines on November 14, leaving 190 people dead, 160 missing, and 120,000 homeless.

1991

Floods in Pakistan killed 24 people in February.

On March 10, floods in Mulanje, Malawi, killed more than 500 people and left 150,000 homeless.

On April 10, a tornado at Sripur, Bangladesh, destroyed a textile mill, killing 60 people and trapping 100 beneath debris.

On April 26, more than 70 tornadoes in Kansas killed 26 people and injured more than 200.

A cyclone with a storm surge producing waves 20 feet (6 meters) high struck coastal islands in Bangladesh on April 30, killing at least 131,000 people and leaving 5,000 fishermen unaccounted for.

A tornado at Tungi, Bangladesh, killed 100 people on May 7.

A tornado at Sirajganj, Bangladesh, killed 13 people on May 9.

May floods in Bangladesh killed more than 100 people and left more than one million homeless.

Floods in China from May to August killed at least 1,800 people.

Flash floods in Jowzjan Province, Afghanistan, killed up to 5,000 people in June.

Floods in Sri Lanka killed 27 people in June.

Floods and landslides in India and Bangladesh killed at least 80 people in July.

Typhoon Amy struck southern China on July 20 and 21, killing at least 35 people.

Prolonged, heavy rain caused a dam to burst near Onesti, Romania, on July 28, killing more than 100 people.

Monsoon rains caused a dike to burst in Maharashtra State, India, on July 30, inundating 52 villages and leaving 475 people dead and 425 missing.

Floods in Cameroon and Chad killed at least 41 people in August.

Hurricane Bob struck the east coast of the United States from August 18 to 20, killing 16 people.

Typhoon Gladys struck South Korea on August 23, bringing 16 inches (40.6 centimeters) of rain to Pusan and Ulsan, killing 72 people, and leaving 2,000 homeless.

Floods in Cambodia killed 100 people in September.

Floods in Bangladesh killed 250 people in September.

Floods in West Bengal, India, from September 7 to 14 killed 40 people.

Floods in Vietnam killed 17 people in September.

Typhoon Mireille, category 4, with winds up to 133 MPH (214 KPH), struck Kyushu and Hokkaido, Japan, on September 27, killing 45 people.

Typhoon Ruth, category 5, with winds up to 143 MPH (230 KPH), struck Luzon, Philippines, on October 27, killing 43 people.

Floods in the Philippines on November 5, caused by Tropical Storm Thelma, left 3,000 people dead.

Floods in southern India killed 125 people in November.

Typhoon Val, category 5, with winds up to 150 MPH (241 KPH), struck Western Samoa from December 6 to 10, leaving 12 people dead and 4,000 homeless.

Floods in Texas killed 15 people on December 21 and 22.

1992

Floods and mudslides in Rio de Janeiro, Brazil, killed 25 people on January 5.

Blizzards in southern Turkey from February 1 to 7 caused avalanches and snowslides that killed 201 people.

In February, floods in Los Angeles and Ventura counties, California, killed eight people and caused $23 million of damage.

Floods in Jiangxi Province, China, killed 29 people in March.

Floods in Tajikstan from May 13 to 15 killed at least 200 people.

Heavy rain caused the Paraguay, Paraná, and Iguaçu rivers to overflow in May and June, causing floods in Argentina, Brazil, and Paraguay that inundated hundreds of towns, killing 28 people and requiring the evacuation of 220,000.

Floods in Fujian and Zhejiang Provinces, China, killed more than 1,000 people in July.

Monsoon rains caused flooding in southern Pakistan in July and August, killing 56 people and leaving thousands homeless.

Hurricane Andrew, category 5, with winds up to 164 MPH (264 KPH), struck the Bahamas, then moved to Florida and Louisiana from August 23 to 26. Homestead and Florida City, Florida, were almost destroyed. In Florida, the hurricane killed 38 people, destroyed 63,000 homes, and caused $20 billion of damage. In Louisiana it left 44,000 people homeless and caused $300 million of damage. Andrew was the most costly hurricane in U.S. history.

Tropical Storm Polly caused a storm surge with 20-foot (60-meter) waves at Tianjin, China, on August 30 and 31, killing 165 people along the southeastern coast and leaving more than five million homeless.

Tsunamis caused by a submarine earthquake struck the western coast of Nicaragua on September 1, killing 105 people and injuring 489.

Flash floods near Gulbahar, Afghanistan, on September 2 destroyed villages and killed up to 3,000 people.

Monsoon rains caused the Indus River to overflow its banks, causing floods in Pakistan and India from September 11 to 16. More than 2,000 people died in Pakistan, and at least 500 in India.

Flash floods in Manila, Philippines, killed 10 people on September 15.

Flash floods in the Ardèche, Vaucluse, and Drôme regions of France on September 22 left 80 people dead and 30 missing.

In October, floods and landslides in Kerala, India, killed 60 people and left thousands homeless.

In November, floods and mudslides in southern India killed 230 people and left thousands homeless.

Floods in the Ukraine killed 17 people in November.

November floods in Albania, caused when the Mati River over-flowed, killed 11 people.

From November 21 to 23, up to 45 tornadoes across 11 states, from Texas to Ohio, killed 25 people.

A cyclone, with heavy rain and snow, struck the northeastern United States on December 10 and 11, killing 17 people and causing $10 million of damage in Atlantic City.

1993

Cyclone Kina, category 4, with winds up to 115 MPH (185 KPH), struck Fiji on January 2 and 3, killing 12 people.

Floods and mudslides in Tijuana, Mexico, and southern California from January 7 to 20 killed 30 people and left 1,000 homeless.

On January 8 in the Sylhet and Sunamganj districts of Bangladesh, a tornado lasting five minutes killed 32 people and injured more than 1,000.

In February, floods in Java, Indonesia, killed 60 people, destroyed many homes, and required about 250,000 people to be evacuated.

February floods in Ecuador caused great damage and an unknown number of deaths.

February floods in Iran killed about 500 people and caused about $1 billion of damage.

From March 12 to 15, a blizzard killed at least 238 people in the eastern United States, as well as 4 in Canada and 3 in Cuba, and caused $1 billion of damage.

A tornado in West Bengal, India, on April 9 destroyed 5 villages and killed 100 people.

On April 26, floods and landslides in Colombia, caused when the Tapartó River overflowed, killed up to 100 people.

Floods and mudslides in Santiago, Chile, killed 11 people on May 3.

From June to August, floods in the midwestern United States, caused when the Missouri and Mississippi rivers overflowed, killed 50 people and caused $12 billion of damage in Illinois, Iowa, Kansas, Minnesota, Missouri, Nebraska, North and South Dakota, and Wisconsin.

Floods in Dhaka, Bangladesh, killed nearly 200 people in June.

In July, monsoon rains caused floods in Himachal Pradesh State, India, in which at least 175 people died.

Hurricane Calvin struck Mexico on July 6 and 7, killing 28 people.

Monsoon rains caused floods in Bangladesh, Nepal, and India in July and August, killing thousands of people.

Floods and landslides in Hunan and Sichuan provinces, China, killed 120 people in July.

In July and August, floods and mudslides in Japan left 40 people dead and 22 missing.

Tropical Storm Bret caused floods and mudslides in Venezuela on August 8; at least 100 people died.

Typhoon Yancy, category 4, with winds up to 130 MPH (209 KPH), struck Kyushu, Japan, in September, killing 41 people.

In September, floods and mudslides in Nicaragua, Honduras, and Mexico killed at least 42 people.

Mudslides in northern Honduras from October 31 to November 2 destroyed 1,000 homes and killed 400 people.

Typhoon Kyle struck Vietnam on November 23, killing at least 45 people.

A cyclone struck southern India in December, killing 47 people.

Hurricane-force winds struck the United Kingdom in December, killing 12 people.

Floods and mudslides in Dabeiba, Colombia, on December 17 left 22 people dead, 35 injured, and several missing.

Mudslides at Oran, Algeria, on December 25 killed 12 people and injured 46.

Typhoon Nell struck the Philippines on December 25 and 26, killing at least 47 people.

Floods in Malaysia killed 14 people in December.

Floods in Belgium, France, Germany, Luxembourg, the Netherlands, and Spain killed at least seven people in December.

1994

Floods and landslides in the Philippines on January 7 left 15 people dead and 30 missing.

February floods in Colombia killed 19 people and destroyed 1,400 homes.

Cyclone Geralda, category 5, with winds up to 220 MPH (354 KPH), struck Madagascar from February 2 to 4, leaving 70 people dead and 500,000 homeless. In the port of Toamasina, 95 percent of the buildings were destroyed.

February floods and mudslides in Peru killed 50 people and left 5,000 homeless.

Tornadoes in Alabama, Georgia, North and South Carolina, and Tennessee killed 42 people on March 27.

A cyclone struck Nampula Province, Mozambique, in March, killing 34 people and leaving 1.5 million homeless.

A cyclone, category 5, with winds up to 180 MPH (289 KPH), struck Bangladesh on May 2, killing 233 people.

Floods in Guangdong and Guangxi Provinces, China, killed up to 400 people in June.

On June 3, earthquakes sent a series of tsunamis onto the east coast of Java, Indonesia. More than 200 people were killed while they slept at Banyuwangi.

In June and July, monsoon rains caused floods in India; about 500 people died.

Floods in the Philippines killed 68 people in July.

A typhoon, category 1, with winds up to 85 MPH (136 KPH), struck Taiwan in August, killing 10 people.

Typhoon Fred struck Zhejiang Province, China, on August 20 and 21, killing about 1,000 people and causing more than $1.1 billion of damage.

On August 26, floods in Baluchistan Province, Pakistan, killed 24 people when their minibus was swept away by the waters.

Floods in Moldova on August 27 and 28 killed at least 50 people.

Floods in Niger killed 40 people in August.

Floods in Algeria on September 23 killed 29 people.

Floods at Houston, Texas, from October 16 to 19 killed 10 people.

Typhoon Teresa struck Luzon, Philippines, on October 23, killing 25 people.

Floods in northern Italy on November 4 and 5 killed about 100 people.

Tropical Storm Gordon struck the Caribbean, Florida, and South Carolina from November 13 to 19, killing 537 people and causing at least $200 million of damage.

A cyclone struck northern Somalia in November, killing 30 people.

1995

In January, violent storms caused floods in much of California. At least 11 people died, and the damage was estimated at $300 million.

In late January and early February, heavy rain and melting snow combined to make the Rhine, Main, Mosel, Meuse, Waal, and Nahe rivers overflow, causing widespread floods in Belgium, France, Germany, and especially the Netherlands. About 30 people died, and damage was estimated at more than $2 billion.

In early March, California suffered further floods, in which at least 12 people died.

On March 27, heavy rain triggered a mudslide in Afghanistan. A village was destroyed, killing 354 people and injuring 64.

In early May, floods and landslides in northern Sumatra, Indonesia, killed at least 55 people and made about 17,500 homeless.

On May 17, a heavy rainstorm and tidal surge combined to kill nearly 100 people in southeastern Bangladesh.

In late May rains washed away a feeding center in Angola, killing 33 people, 25 of them children.

On June 3 near Dimmitt, Texas, a tornado lasting about 20 minutes destroyed a home, lifted automobiles from the ground, and then removed a section of asphalt more than 475 square yards (397 square meters) in area from a highway, dropping it about 220 yards (201 meters) from its original position. The tornado produced winds of more than 155 MPH (249 KPH) and a downdraft of more than 56 MPH (90 KPH) at the eye. This was the first tornado to be studied using new radar equipment that revealed more detail of its internal structure than had been observed previously.

In June, pre-monsoon rains caused widespread flooding and land-slides in Bangladesh and Nepal. At least 50 people were killed in Bangladesh and 60 in Nepal.

In June and July heavy rains caused flooding in Hunan, Hubei, and Jiangxi provinces, China. At least 1,200 people died, and about 5.6 million were stranded. About 900,000 homes were destroyed and 4 million damaged, requiring the relocation of 1.3 million people.

On July 13 flash floods triggered a mudslide at Senirkent, Turkey. About 200 homes were destroyed, and at least 50 people died.

In mid-July more than 150 people died in widespread floods in Bangladesh, and nearly 600 people died in floods in Pakistan.

In mid-July prolonged, heavy rain caused a landslide that descended on a village in southwestern China during the night and buried it, when everyone was asleep, killing 26 people.

In mid-July, Typhoon Faye struck South Korea. At least 16 people died, and 25 were missing.

On August 17, heavy rain caused flash floods in the Atlas Mountains, Morocco, killing more than 230 people and leaving about 500 missing.

In early September, monsoon rains caused the deaths of at least 40 people in northern India.

Further flooding in Morocco in early September killed 31 people.

Hurricane Luis, category 4, with winds gusting to more than 140 MPH (225 KPH), struck Puerto Rico and the U.S. Virgin Islands from September 4 to 6. At least 15 people were killed.

On September 11, the World Health Organization (WHO) announced it had provided $100,000 to meet the immediate health needs of flood victims in North Korea. Flooding began in late July, rendering 100,000 families homeless and affecting five million people, nearly 25 percent of the population.

Hurricane Ismael struck Mexico on September 14, killing at least 107 people, many of them fishermen at sea, in the northwestern Pacific states.

In mid-September floods affected 52 of the 76 provinces of Thailand. At least 62 people died.

Hurricane Marilyn struck the U.S. Virgin Islands and Puerto Rico from September 15 to 16, with winds of more than 100 MPH (160 KPH), leaving 9 people dead, 100 injured or missing, and destroying 80 percent of the houses on St. Thomas.

In late September and early October, five days of heavy rain caused widespread flooding in Bangladesh. More than 100 people died, and more than one million were trapped in their homes.

On October 1, Tropical Storm Sybil struck the Philippines, causing damage in 29 provinces and 27 cities. It triggered floods, landslides, and volcanic mudflows. More than 100 people were killed, and 100 were left missing, and feared killed.

Hurricane Opal, category 4, with winds of 150 MPH (241 KPH), formed over the Yucatán Peninsula, Mexico, on September 27. By the end of the month Opal was classed as a tropical storm, but became a hurricane by October 2. It killed 50 people in Guatemala and Mexico and 13 in the United States, although it had weakened to category 3 by the time it reached Florida on October 4. It then moved into North Carolina, Georgia, and Alabama. The damage Opal caused was due mainly to a storm surge producing 12-foot (3.6-meter) storm-surge waves and breaking waves. Damage was estimated at more than $2 billion.

Hurricane Roxanne, category 3 with 115-MPH (185-KPH) winds, struck the island of Cozumel, off the Mexican coast, in October, killing 14 people in Mexico and driving tens of thousands from their homes.

Following days of fierce storms and blizzards, before dawn on October 26 a snowslide engulfed 19 homes and killed 20 people in the fishing village of Flateyri, Iceland.

Tropical Storm Zack struck the Philippines in late October, causing severe flooding, capsizing a ship sailing between islands, and killing at least 59 people. In all, at least 100 people died, and 60,000 had to leave their homes.

Typhoon Angela, category 4, struck the eastern Philippines on November 3 with winds of 140 MPH (225 KPH). It killed more than 700 people, destroyed 15,000 homes, and left more than 200,000 people homeless. Damage to crops, roads, and bridges amounted to $77 million.

From November 11 to 12, heavy snow triggered snowslides and mudflows in Nepal, killing at least 49 people.

On December 25, flash floods caused by prolonged, heavy rain killed at least 130 people in Natal Province, South Africa.

In late December, extreme cold and blizzards affected Europe and Asia from Britain to Kazakhstan and Bangladesh. More than 350 people died, many of them in Moscow, where people froze while drunk.

In late December, about 60 people died in floods in Brazil caused by heavy rain.

1996

From January 6 to 9 the worst blizzards in 70 years swept the eastern United States. At least 23 people died as snow driven by 25–35-MPH (40–56-KPH) winds affected Alabama, Indiana, Kentucky, Maryland, Massachusetts, New Jersey, New York, North Carolina, Ohio, Pennsylvania, Rhode Island, Virginia, Washington, D.C., and West Virginia. States of emergency were declared in Kentucky, Maryland, New Jersey, New York City, Pennsylvania, Virginia, and West Virginia. President Clinton designated nine states as disaster areas. No mail was delivered in New York City on January 9, and the United Nations building was closed.

On May 13, a tornado with 125-MPH (201-KPH) winds killed more than 440 people and injured more than 32,000 in Bangladesh. Lasting less than half an hour, it destroyed 80 villages, striking hardest in the Tangail district, 45 miles (72 kilometers) north of Dhaka. Of those killed, at least 120 were in the village of Bashail, some of them students at a boarding school that collapsed; 55 people died in Rampur; and there were deaths in Gopalpur, Kalihati, Shafiur, Mirzapur, and Ghatail.

On June 3, a 60-acre (24-hectare) brushfire in Alaska turned into a huge wildfire when it ignited forest-floor moss that had dried as a result of drought; the wildfire was driven by 25-MPH (40-KPH) winds. Firefighters suspected the fire had been started by fireworks. By June 6 it had grown into 60 separate wildfires covering more than 40,000 acres (16,000 hectares); these were still burning at the end of the month.

On June 16, a cyclone and heavy rainfall caused extensive flooding in the southern Indian states of Andhra Pradesh, Tamil Nadu, and Karnataka. Telephone and power lines were brought down, thousands of people were driven from their homes, and at least 100 died. More than 190 people, most of them fishermen, were reported missing.

On June 19, storms brought severe flooding in Tuscany, Italy. Mudslides caused rivers to burst their banks, and bridges were swept away. Fornovalasco and Cardoso were the worst affected villages, inundated by the River Vezza. More than 1,000 inhabitants were evacuated from the two villages on June 20. The main rail line between Rome and Genoa was blocked by a mudslide. By June 21, 11 bodies had been recovered and 30 people were missing.

In late June, drought led to wildfires in Nevada, New Mexico, Utah, and Arizona. The Nevada fire, started by boys playing with gasoline 60 miles (96 kilometers) south of Reno, burned about 4,000 acres (1,600 hectares). On June 23, between 3,000 and 4,000 people were evacuated from their homes, but were allowed to return the following day. In Arizona, the fires burned more than 31,500 acres (12,700 hectares) of forest. In Utah, 30,000 acres (12,100 hectares) were burned; 360 acres (145 hectares) in the Jemez Mountains, about 45 miles (72 kilometers) northwest of Albuquerque, New Mexico, in the Santa Fe National Forest were destroyed. By July, as the drought continued, drinking water was becoming scarce in parts of Texas. On July 27, the Nevada fires were brought under control with the help of steady drizzle, but lightning ignited more fires in Utah, damaging 8,000 acres (3,200 hectares) of the Dixie National Forest; a fire in Idaho burned 13,000 acres (5,200 hectares) of brush; and in Colorado a 5,340-acre (2,161-hectare) fire was being brought under control. By late August, 84,000 fires had burned a total of 4.8 million acres (1.94 million hectares) in nine western states. More fires began in Oregon on August 26, adding to others already burning and affecting 100,000 acres (40,400 hectares) in the state; fires were still raging in California, Washington, Idaho, Utah, Nevada, Montana, and Wyoming. By the end of August nearly 20,000 firefighters were battling with blazes across the western states.

At the end of June, the Orinoco River burst its banks when its level rose in some places by 10 feet (3 meters) following two weeks of rain. More than 10,000 acres (4,000 hectares) of lands cultivated by Amazonian tribes was flooded, and on June 28 the governor of Amazonas State, Venezuela, declared a state of emergency.

Two weeks of monsoon rains in late June and early July caused the Ganges and Brahmaputra rivers to burst their banks, flooding nearly one-third of Bangladesh and affecting 3 million people. On July 7 the Jamuna River inundated 50 villages, displacing nearly 70,000 people in Sirajganj, 65 miles (104 kilometers) northwest of Dhaka. Lack of safe drinking water led to outbreaks of typhoid and diarrhea, affecting more than 1,000 people. The rains and flooding continued for more than a month, killing at least 115 people, and nearly 6 million lost their homes and crops. The same monsoon rains caused 82 deaths in Nepal and 78 in India.

Heavy rain, beginning in late June and continuing into August, caused widespread flooding in 11 provinces of central and southern China. In Lin'an and Tonglu, southwest of Hangzhou, the capital of Zhejiang Province, the water was 20 feet (6 meters) deep on July 8. More than 1,500 people died, and the floods were estimated to have destroyed at least 2.5 million acres (1 million hectares) of crops and injured or damaged the property of 20 million people. By mid-July the Yangtze and other rivers in central China were still rising and Dongting Lake overflowed, flooding 39 counties and cities in Hunan, in places to a depth of 20 feet (6 meters). About eight million troops, police, reservists, officials, and military academy students were mobilized to rescue more than 650,000 people. The cost of the floods was estimated at about $12 billion. The floods started receding on July 25, but returned in early August. Mongolia also suffered floods that killed 41 people, and the capital, Ulan Bator, was inundated when two rivers overflowed.

Starting on July 6, snowstorms affected large areas of eastern South Africa, leaving thousands of travelers and many hikers stranded. More than 3,000 people were trapped at Harrismith, where 6-foot (1.8-meter) drifts blocked a pass. By July 10, the death toll had reached 44, including two boys in Lesotho. By July 11, 96 hikers had been rescued by helicopter from the Drakensberg Mountains, where they had been stranded since July 6, and roads reopened. South of Johannesburg, 91 people were rescued. In villages in the Maluti Mountains, Lesotho, more than 100,000 people were cut off. Temperatures fell to 10°F (-12°C) at Lady Grey, Free State Province, south of Johannesburg. The Weather Bureau in Pretoria said this snowfall was the heaviest and most sustained since 1962; in some parts of Free State Province, it was the first snowfall in 50 years.

On July 7, two people died in flooding in the area around Lake Maggiore, in northern Italy, one a woman of 67 in a landslide near Omegna, the other a German canoeist who was swept away near Cannobio. By July 8, the main road into Omegna was under water and strewn with debris. The floods were caused by heavy storms.

On July 8, Hurricane Bertha reached the U.S. Virgin Islands, with torrential rain and winds up to 103 MPH (165.7 KPH); Bertha crossed St. Thomas, but came only within 45 miles (72 kilometers) of Puerto Rico, where winds reached 85 MPH (136 KPH) and 5–8 inches (12.7–20.32 centimeters) of rain fell, triggering flash floods and mudslides. Hurricane Bertha crossed the British Virgin Islands, where roofs were torn from buildings, power lines were brought down, and buildings were flooded by heavy rain, then moved on to St. Kitts and Nevis, and Anguilla. On July 10, Bertha reached the Bahamas, with 20-foot (6-meter) waves and 100-MPH (160-KPH) winds; on July 12, the hurricane reached the Carolina coast and moved north, eventually to Delaware and New Jersey. By July 11, Bertha had caused six deaths in Puerto Rico and the Virgin Islands, and one in Florida. Radio calls were received from a Venezuelan ship adrift off Puerto Rico near the eye of the storm, with 42 people on board and only 14 lifejackets, and the U.S. Coast Guard conducted a search. Bertha was initially classified as a category 1 hurricane, but was later upgraded to category 2 and on July 9, as its winds reached 115 MPH (185 KPH), category 3. The hurricane was large, with a diameter of 460 miles (740 kilometers), and its winds reached hurricane force 115 miles (185 kilometers) from the eye.

On July 12, a severe storm caused flash flooding at Buffalo Creek, Colorado, destroying two roads and a bridge, as well as several buildings.

On July 13, monsoon rains caused floods in northern Bengal and a landslide in West Bengal State. The flood swept 14 people to their deaths and left nearly 100,000 marooned, and 37 died in the landslide.

By mid-July, monsoon rains had caused floods that inundated 60 villages in Assam State, India. The state government set up 120 relief camps to accommodate 1.5 million people.

In mid-July, a tornado destroyed electrical and communications equipment and several hundred houses in the cities of Kiangyan and Taixing, in Kiangsu Province, China. Many farmers were killed or injured by collapsing houses. The tornado killed 21 people and injured more than 200.

On July 18, Typhoon Eve crossed the Satsuma peninsula on the southern tip of Kyushu, the southernmost island of Japan, with maximum wind speeds of 119 MPH (191 KPH); later the typhoon weakened.

After two days in which 11 inches (27.9 centimeters) of rain fell —more than what ordinarily falls in the whole of July—on July 21, rivers burst, causing widespread flooding in the Saguenay River region of southern Québec, about 200 miles (320 kilometers) north of Montréal. At least eight people died. About 100 houses were destroyed, as well as several other buildings in La Baie, where 12,000 residents had to leave their homes and were not allowed to return for some days, for fear that nearby dams might burst. About 3,000 were accommodated in tents at a military base at Bagotville, and two tents were also provided for 40 cats and dogs. The Québec government created a $200 million relief fund, but that was believed to be a conservative estimate of the cost of the flooding.

On July 23, Tropical Storm Frankie moved from the Gulf of Tonkin to the Red River delta, Vietnam, bringing winds of 56 MPH (90.1 KPH) and 4.5–8 inches (11.4–20.3 centimeters) of rain. After two days the storm abated, leaving 41 people dead and many missing.

On July 25, Typhoon Gloria, with winds gusting to 106 MPH (170.5 KPH), caused damage in the Philippines and caused at least 30 deaths. It was the sixth major storm to strike the islands in 1996.

Gloria, downgraded to a tropical storm, reached Taiwan and the southeast coast of China on July 26, killing three people.

Five days of heavy rain in late July delivered up to 20 inches (170.5 centimeters) of rain on rice-growing areas in North and South Korea; some parts of North Korea received nearly 29 inches (73.7 centimeters). UN officials reported that at least 230 people died in North Korea. Rivers flooded, there were landslides, and in South Korea 64 people died, 44 of them soldiers buried by mud inside their barracks and posts. The South Korean border towns of Yonchon and Munsan were flooded to roof level, and about 50,000 people had to leave their homes in an area north of Seoul. Damage was estimated at $600 million. In North Korea the floods followed heavy flooding in 1995 and destroyed 20 percent of the country's annual food production, threatening famine.

Typhoon Herb, with winds of 121 MPH (194.7 KPH), reached Taiwan on July 31, bringing 44 inches (111.8 centimeters) of rain in 24 hours at Mount Ali, southern Taiwan, flooding thousands of homes. The typhoon killed at least 41 people, and 34 were left missing. It was the worst storm to strike Taiwan in 30 years. Herb then crossed the Taiwan Strait to China, reaching Pingtan, in Fujian Province, on August 1, with winds of 87 MPH (140 KPH).

On July 7, monsoon rains triggered landslides in northeastern Nepal, in an area 55 miles (88 kilometers) northeast of Katmandu. Dozens of homes were destroyed in the village of Jhagraku, and at least 40 people died.

On July 7, after two days of heavy rain, a campsite in the Spanish Pyrenees was destroyed by flood water, mud, and rocks. At least 100 people were killed.

On August 14, Typhoon Kirk crossed southwestern Honshu, Japan, with winds of 130 MPH (209 KPH) and up to 12 inches (30.5 centimeters) of rain. On August 15, it returned and crossed the northern part of the island, but with winds of only 56 MPH (90.1 KPH). Kirk then moved to northeastern China, causing floods that inundated 845 villages along the Yellow River.

By mid-August, two weeks of heavy rain in the far east of Russia had flooded eight towns and 126 villages along the Ussuri, Ussurka, and Malinovka rivers, killing four people. Roads, bridges, and crops were destroyed, power and telephone lines brought down, and 3,500 buildings damaged. The damage was estimated to cost $140 million.

On August 20, Tropical Storm Dolly, with winds of 30 MPH (48 KPH), strengthened to hurricane force when its windspeed reached 75 MPH (120 KPH) as it reached Punta Herrero on the Caribbean coast of Mexico. It then weakened, but regained hurricane force as it moved back over the sea and headed toward northeastern Mexico and Texas.

In August, North Vietnam suffered severe storms, with a waterspout offshore from Hau Loc on August 14. The storm destroyed many fishing boats and, together with flooding due to heavy rain, killed 53 people. More than 1,000 homes were destroyed by flash floods and mudslides, leaving 10,000 people homeless near the border with Laos; on August 21, Hanoi was threatened with flooding as the level of the Red River rose 1.6 feet (48 centimeters) above the flood-warning mark. The following day it burst its banks at a dike about 30 miles (48 kilometers) northwest of Hanoi, causing 80,000 people to flee their homes in the city, which was under several feet of water. Tropical Storm Niki, downgraded from a typhoon, struck just south of Haiphong on August 23, sinking four cargo ships and leaving one person dead.

On September 1, Hurricane Edouard headed toward Cape Cod, and a state of emergency was declared in Massachusetts. Hurricane winds had by then weakened from 140 MPH (225 KPH) to 100 MPH (160 KPH). Edouard had already battered New Jersey, killing two people. Hurricane Fran was then 290 miles (466 kilometers) northeast of Puerto Rico in the Caribbean, with winds of 80 MPH (128 KPH) and expected to strengthen, and Hurricane Gustav was west-southwest of the Cape Verde Islands.

In late August, prolonged, heavy rain caused widespread flooding in northern Bangladesh. The Padma River overflowed, and some low-lying outskirts of the city of Rajshahi (population two million) were under 2 feet (60 centimeters) of water. Rice crops were damaged in Rajshahi, Chapainawabgani, Dinajpur, and Sherpur. At least half a million people were affected, 100,000 had to leave their homes, and 20 died.

On September 2, following two hours of heavy rain, flash floods destroyed rail lines and bridges in and around al-Geili, north of Khartoum, Sudan. Thousands of people were made homeless, and 15 died.

On September 6, Hurricane Fran, category 3 with winds of 115 MPH (185 KPH) gusting to 125 MPH (201 KPH) extending to 145 miles (233 kilometers) from the eye, passed Cape Fear, North Carolina, shortly before 8:00 p.m. heading north. By midnight winds had dropped to 100 MPH (160 KPH) and the eye had disappeared, but the hurricane produced tornadoes and a storm surge of up to 12 feet (3.6 meters) and, in some places, 16 feet (4.8 meters). More than 500,000 people were ordered to evacuate coastal areas of South Carolina, and evacuation was also ordered in all or part of eight counties in North Carolina. By the following day Fran had swept through North and South Carolina, Virginia, and West Virginia, causing 34 deaths and damage that was preliminarily estimated at $625 million, but was expected to rise. President Clinton declared major disasters in North Carolina and Virginia, and state governors declared emergencies in Virginia, West Virginia, and North Carolina. The flooding caused by the hurricane was followed by a second wave of heavy rain in eastern North Carolina, causing the Neuse River to rise about 13 feet (3.9 meters). Roads were closed, creeks and rivers overflowed, and 1,000 people were evacuated from their homes in Kinston. Then, on September 16, a wave of heavy rain and tornadoes swept the southeastern part of the state, causing more damage, some of it about a mile (1.6 kilometers) from Kinston.

On September 7, Tropical Storm Hortense delivered heavy rain over Martinique, causing flooding and bringing down power lines. It then headed for the British Virgin Islands and Puerto Rico, with forecasts of up to 10 inches (25.4 centimeters) of rain and winds of 60 MPH (96 KPH) extending 100 miles (160 kilometers) from the center. By September 10, Hortense had strengthened to a category 1 hurricane; it delivered up to 20 inches (50.8 centimeters) of rain on Puerto Rico, causing flash floods, mudslides, and damage estimated to cost $155 million, then the hurricane moved to the Dominican Republic. It killed 16 people and left dozens missing there and in Puerto Rico, then headed for the Bahamas and Turks and Caicos Islands.

On September 10, Typhoon Sally passed Hong Kong and crossed the coast of Guangdong, China, with winds of up to 108 MPH (173.8 KPH). More than 130 people were killed, thousands were injured, nearly 400,000 homes were destroyed, and more than 30 fishing boats were sunk.

In September, tropical storms caused floods in central Vietnam in which 17 people died and 114,000 had to leave their homes.

On September 22, Typhoon Violet crossed Japan, with winds of 78 MPH (125 KPH) gusting to 116 MPH (186 KPH) and 10.4 inches (26.4 centimeters) of rain in Tokyo. By September 23, Violet had weakened to a tropical storm and was moving into the Pacific. At least seven people were killed, most in the Tokyo area. In Honshu, Violet caused about 200 landslides, destroyed more than 80 homes, and flooded more than 3,000.

In late September, Typhoon Willie struck the Chinese island of Hainan, killing at least 38 people and leaving 96 missing. In the capital, Haikou, 70 percent of the streets were flooded. Some parts of the island had more than 15 inches (38.1 centimeters) of rain. Three of the fatalities were in the county of Wenchang, where the typhoon breached breakwaters and swept away 53 fishing boats and 43 houses. The Niandu River inundated 95,000 acres (38,000 hectares) of farmland around the city of Qiongshan.

More than a week of heavy rain in September caused widespread flooding in Bosnia. A state of emergency was declared in several regions, and roads, power, and telephone lines were damaged.

On September 25, Tropical Storm Isidore formed in the eastern Atlantic and strengthened. Isidore was traveling west-northwest at about 21 MPH (34 KPH), with winds of up to 65 MPH (104 KPH), and was expected to strengthen further.

On September 28, Typhoon Zane crossed Taiwan, triggering mudslides and killing two people. Zane then moved to Okinawa.

On September 29, the Cambodian government declared a state of emergency after floods caused by heavy monsoon rains inundated 100 homes and damaged nearly 30,000 acres (12,000 hectares) of farmland. The following day, flood waters entered Phnom Penh. At least 11 people died, and the floods affected 3 million. Across the border in Laos, the floods and landslides were the worst in living memory. At least 30 people died, and rice fields and homes were devastated. By October 2, the Mekong River, Vietnam, was flooding along the Cambodian border. Levels continued to rise; by October 7, 21 people had died in Vietnam, 200,000 homes had been flooded, and more than 24,700 acres (10,000 hectares) of crops had been inundated.

Heavy rains brought flooding to Matamoros, Mexico, starting on October 4. By the following day, the city streets were under 3 feet (90 centimeters) of water. More than 1,500 people were made homeless in the area, and 2 died.

Tropical Storm Josephine reached the Florida coast on the night of October 7, bringing winds of 70 MPH (112 KPH) and up to 5 inches (12.7 centimeters) of rain, as well as triggering tornadoes. The storm produced a storm surge of 6–9 feet (1.8–2.7 meters).

1997

A tornado moving at 260 MPH (418 KPH) leveled a five-mile-long (800 meters), half-mile-wide (8 kilometers) area of Jarrell, Texas on May 29, killing 30, injuring scores of others, and destroying about 50 houses.

On July 5, torrential rains in northeastern Iran caused flash floods that drowned 11 people, washed away hundreds of homes, and damaged pastures and farmland.

July floods caused by a week of heavy rainfall devastated southern China, killing 56 people, stranding thousands more, and inflicting more than $220 million in property damage.

In Villa Angel Flores, Mexico, a small tornado swept up toads from a nearby pond and dropped them like rain on the town on July 8.

A chronology of discovery

c. 340 B.C.

Aristotle wrote *Meteorologica*, the oldest description of weather phenomena known, and quite possibly the first. It consisted of observations of weather and attempts to explain how they happen, and also gave us our word *meteorology* (literally, "discourse on lofty matters").

Aristotle (384–322 B.C.) was a Macedonian who traveled to Athens when he was 17 and studied under Plato. After Plato's death in 347 B.C., Aristotle left and traveled widely, spending several years as tutor to a teenager later known as Alexander the Great. Aristotle wrote on many topics, but was especially renowned for his scientific work. He emphasized the importance of learning from nature by observation and made many meticulously detailed studies.

140–131 B.C.

In China, Han Ying wrote *Moral Discourses Illustrating the Han Text of the "Book of Songs."* This contained the first known reference to the hexagonal (six-sided) structure of snowflakes.

1st century B.C.

The octagonal Tower of Winds was built in Athens. It was topped by a wind vane representing Triton, a sea god, which pointed toward one or another of eight demigods depicted at the top of its walls. Each of the demigods was associated with particular weather conditions, so the tower indicated the kind of weather people should expect according to the direction of the wind. This was possibly the world's first attempt at weather forecasting.

c. 55 B.C.

Lucretius proposed that thunder is the sound of great clouds crashing against one another. He was mistaken, but may have been the first person to note that thunder occurs only in the presence of large, solid-looking clouds.

Titus Lucretius Carus (*c.* 94–55 B.C.) was a Roman philosopher and poet. Only one of his works has survived, *De Rerum Natura* ("On the Nature of Things"). He died before completing his final revisions to it. In *De Rerum Natura*, Lucretius described in verse form the ideas of the Greek philosopher Epicurus (*c.* 342–270 B.C.), holding that all matter is composed of arrangements of atoms of varying sizes.

1st century A.D.

Hero of Alexandria wrote a book, *Pneumatica*, in which he demonstrated that air is a substance. He did this by showing that if a vessel is filled with air, water will not enter it unless the air is allowed to escape. It was already known that air can be compressed. Hero argued that this proved air must be made of tiny particles with space between them, so compression made the particles move closer together.

Hero was a mathematician, and his writings suggest he taught what we would now call physics. He described many devices, some of which he probably invented. These included coin-operated machines and an "engine" consisting of a hollow sphere containing water with two tubes emerging from it. The tubes were bent and pointed in opposite directions. When the sphere was heated, the water boiled and steam discharging from the tubes made the device spin rapidly.

1555

In Rome, Olaus Magnus published a book on natural history containing the first European depictions of ice crystals and snowflakes.

Olaus Magnus was the Latinized name of Olaf Mansson (1490–1557), a Swedish priest born at Linköping. In 1523 he traveled to Rome and later lived in Danzig and then Italy with his brother, Archbishop Johannes Magnus (1488–1544). After his brother's death, Mansson was made Archbishop of Sweden. He compiled a detailed map of Scandinavia and wrote a history of its peoples that became very famous. Translated into English as *History of the Goths, Swedes and Vandals* (1658), it provided the description on which other Europeans continued for many years to base their ideas about the Scandinavian people and countryside.

1586

Simon Stevinus showed that the pressure a liquid exerts on a surface depends on the height of the liquid above the surface and the area of the surface on which it presses, but does not depend on the shape of the vessel containing it.

Simon Stevinus (1548–1620) was a Flemish mathematician, born in Bruges. He died in either The Hague or Leiden. While serving as a quartermaster in the Dutch army, he devised a system of sluices in the dikes protecting the reclaimed polders (fields below sea level). He also introduced decimal fractions (that is, representing ½ as 0.5, for example) to mathematics. In the same year he made his discovery about water, he performed an experiment in which he dropped two different weights at the same time and found they reached the ground simultaneously. This experiment is usually attributed to his younger contemporary, Galileo.

1591

Thomas Harriot (or Hariot) noted that snowflakes are either six-sided or six-pointed. He did not publish his observation, however.

Thomas Harriot (1560–1621) was an English mathematician, born and educated at Oxford. He was scientific adviser to Sir Walter Raleigh on his expedition to Roanoke, Virginia, in 1585–86, about which he wrote *A Brief and True Report*. Harriot had a special interest in astronomy, and was one of several observers who discovered sunspots at about the same time in 1611, but he also made important contributions to the development of algebra. He died in London.

Figure 1: *Galileo Galilei* (The Free Library of Philadelphia)

1593

Galileo invented a "thermoscope" to measure temperature. It was based on the expansion and contraction of a gas in response to changing temperatures and was highly inaccurate. Nevertheless, this was the first attempt at making a thermometer, and the device remained in use for about 10 years.

Galileo Galilei (1564–1642), always known by his first name, was one of the greatest scientists who ever lived. He was born in Pisa, Italy, the son of a mathematician, who sent him to study medicine. In those days (and still!) a physician earned a great deal more than a mathematician. While a student, Galileo heard a lecture on geometry and persuaded his father to allow him to study mathematics and science instead of medicine.

Whereas other scientists of his day were prepared to observe natural phenomena, Galileo tried to measure them. He believed that by attaching quantities to phenomena, mathematical relationships would appear that would allow these phenomena to be described simply and in ways applicable to a range of similar occurrences. This was his principal scientific innovation, from which his many achievements in astronomy and physics sprang.

The last nine years of his life, from December 1633 to 1642, were spent under house arrest at his small estate at Arcetri, near Florence, following his conviction for heresy by the Inquisition. The story of his trial and the events leading up to it is complicated, but the matter centered on his book *Dialogo dei Massimi Sistemi* ("Dialogue on the Two Chief World Systems"), in which two fictional characters debate the astronomical systems of Ptolemy (holding that the Earth is at the center of the universe) and Copernicus (holding that the Sun is at the center of the universe). Galileo made it clear that the Copernican system was superior, which it was alleged he had been expressly forbidden to do, the church having declared the Copernican view false. Galileo died at Arcetri.

1611

Johannes Kepler published *A New Year's Gift, or On the Six-cornered Snowflake*, in which he described snowflakes.

Johannes Kepler (1571–1630) was a German astronomer, born at Weil, in Württemberg, the son of a professional soldier. In 1594 he became a professor of mathematics at the University of Graz, but was forced to leave in 1600 because of religious persecution. He and his wife moved to Prague, where he took up the post of assistant to the Danish astronomer Tycho Brahe (1546–1601). After Brahe's death in 1602, Kepler had access to the observations Brahe had been accumulating for many years. From these, Kepler calculated that planetary orbits are elliptical, with the Sun at one focus; that if a line is drawn from any planet to the Sun, the area it sweeps out in any given time is always the same; and that the time taken

Figure 2: *Ferdinand II and his wife, Vittoria della Rovere* (The Free Library of Philadelphia)

for a planet to complete one orbit is proportional to the cube of its mean distance from the Sun. These are known as Kepler's three laws of planetary motion. He was also an eminent mathematician, whose studies prepared the way for the infinitesimal calculus. In 1628 he and his family moved to Silesia. He died at Regensburg, Bavaria.

1641
Ferdinand II, the grand duke of Tuscany (1610–70), invented a thermometer in the form of a tube containing liquid and sealed at one end. Also see 1654.

1643
Torricelli invented the mercury barometer by sealing one end of a glass tube, filling the tube with mercury, then inverting the tube with its open end immersed in a bath of mercury. He found the

mercury fell to a height of about 30 inches (76.2 centimeters). This is the reading representing average sea-level atmospheric pressure, nowadays described as 1013.2 millibars (mb).

Evangelista Torricelli (1608–47) was born at Faenza, Italy, and studied mathematics in Rome. In 1638 he read works by Galileo Galilei (1564–1642). Galileo read a book by Torricelli on mechanics and invited him to Florence. Torricelli became an assistant to Galileo, who was then blind, for the last three months of Galileo's life. In making his barometer, Torricelli also produced the first artificially made vacuum, the space above the mercury in the tube being quite empty except for a small amount of mercury vapor.

Figure 3: *Blaise Pascal* (The Free Library of Philadelphia)

1646

The French physicist Blaise Pascal demonstrated that with increasing altitude, the pressure exerted by the weight of overlying air decreases. He suggested that the atmosphere is like an ocean, so climbing to greater heights is like ascending toward the surface. (In fact, the atmosphere has no clearly defined surface resembling the surface of the sea.)

Pascal was chronically sick (with indigestion and insomnia) and unable to climb a mountain himself; he asked his brother-in-law Florin Périer to climb the Puy de Dôme, an extinct volcano 4,800 feet (1,464 meters) high in the Auvergnes region of France, not far from Clermont-Ferrand, where Pascal was born. Périer carried two barometers and found that when he reached the summit, the mercury had fallen 3 inches (7.62 centimeters). This proved that atmospheric pressure decreases with height.

Blaise Pascal (1623–62) was famous as a mathematician and philosopher as well as a physicist. When he was 19 he invented a calculating machine that performed additions and subtractions. The computer programming language Pascal is named for him, in recognition of his invention of this forerunner of the computer. Pascal was one of the founders of the theory of probability, which forms the basis of statistics. In recognition of his work on the effects of pressure on liquids and gases, the international unit of pressure or stress is called the pascal (Pa), equal to a force of one newton per square meter (1 mb = 100 Pa; 1 lb ft^2 = 47.88 Pa). In the year of the climb up the Puy de Dôme, Pascal turned to the Jansenist sect of Roman Catholicism. From that time he became increasingly devout and devoted the remaining years of his life to meditation and writing on religion.

1654

Ferdinand II, grand duke of Tuscany, improved on his thermometer (see 1641), providing the model that would lead in 1714 to the mercury thermometer invented by Gabriel Fahrenheit.

1660

Robert Boyle published *New Experiments Physico-Mechanical Touching the Spring of Air and Its Effects*, in which he reported that not only can air be compressed, but the amount by which it is compressed varies with pressure in a simple way: pv = a constant, where p is pressure and v is volume. This is now known in Britain and North America as Boyle's law, but in France the discovery is attributed to Edmé Mariotte (1620–84) and known as Mariotte's law. The most important conclusion from Boyle's experiment was that since air is compressible, it must consist of individual particles separated by empty space, an idea first expressed by Hero of Alexandria.

Robert Boyle (1627–91), the Irish chemist and physicist, was the 14th child (and 7th son) of the Earl of Cork and was born at Lismore Castle, in Ireland. He was an infant prodigy; by the age of 14, he was studying the work of Galileo. Later he did much to transform alchemy into chemistry, then separated chemistry from medicine and established it as a scientific discipline in its own right. He was a founding member of the Royal Society of London, and in 1680 was elected its president, but declined because of his scruples about taking the necessary oath. He was a devout man who wrote essays on religion and financed missionary work in Asia. He died in London.

1665

Robert Hooke published *Micrographia*, which included illustrations of snowflakes he had observed under the microscope and descriptions of their crystal structures.

Robert Hooke (1635–1703) was an English physicist and in his day was unrivaled as an inventor of instruments. He also improved instruments that already existed, including the barometer; by showing that the height of the mercury changed before a storm, he suggested the possibility of using a barometer to forecast the weather. It was he who first labeled a barometer with the words "change," "rain," "much rain," "stormy," "fair," "set fair," and "very dry." He was fascinated by microscopy and devised a compound microscope (he did not invent it). As well as describing snowflakes, *Micrographia* contained his beautiful drawings of insects, fish scales, feathers, and the structure of cork, which he found contained countless tiny holes that he called "cells," a word that has remained in use.

Hooke was born at Freshwater, Isle of Wight, and died in London.

1686

Edmund Halley proposed the air is heated more strongly at the equator than elsewhere. The warmed air rises and is replaced at the surface by cooler air flowing toward the equator. This was the first attempt to explain the trade winds.

Edmund Halley (1656–1742) was known mainly as an astronomer. After the "Great Comet" was seen in 1680, he calculated its orbit (and those of 23 other comets) and predicted it would return in 1758. The comet now bears his name. In 1720 he was appointed Astronomer Royal. His interests were very wide, however, and his explanation of the trade winds formed part of the first-ever map of surface winds over the whole world. He also found a relationship between height and air pressure. Halley was born in London and died in Greenwich, England.

Figure 4: *Edmund Halley* (The Free Library of Philadelphia)

1687

Guillaume Amontons invented the hygrometer to measure humidity. He also made a barometer that did not use mercury. This instrument could be used at sea, where the movement of the ship caused the mercury level to rise and fall, making readings inaccurate.

Guillaume Amontons (1663–1705) was born and died in Paris. For most of his life he was deaf, but far from regarding this as a disability, he believed it allowed him to concentrate more intently on the scientific work that really interested him.

1714

Fahrenheit invented the mercury thermometer. Alcohol and alcohol-water thermometers already existed, but were inaccurate and unreliable, although Fahrenheit improved them, and the low boiling point of alcohol meant they could not be used to measure higher temperatures. He also devised for his thermometer the scale that still bears his name. This took as zero the lowest temperature that can be reached with a mixture of ice and salt. He marked 96 divisions above that (first 12 divisions, then each of these subdivided into 8) to reach the blood temperature of a healthy person. On this scale, the freezing point of pure water falls at 32°F (0°C). Fahrenheit extended it to the boiling point of pure water, adjusting his scale to make this fall at precisely 212°F (100°C), the effect of the adjustment being to alter the human body temperature to 98.6°F (37°C). His thermometer and its scale quickly became popular, and the scale remains in use in some countries, although it has been superseded by the Kelvin scale for scientific work. Having 180 divisions between the freezing and boiling temperatures of water, the Fahrenheit scale permits finer whole-number measurements than the Celsius scale, which has only 100 divisions.

Gabriel Daniel Fahrenheit (1686–1736) was born in Danzig, Germany (now Gdansk, Poland) and moved to Amsterdam as a young man. He earned his living in the Netherlands as a glassblower and maker of meteorological instruments. He died in The Hague.

1735

George Hadley proposed that the rotation of the Earth causes winds flowing toward the equator to be swung, so they blow from an easterly direction, northeast in the northern hemisphere and southeast in the southern. Hadley believed air warmed at the equator rises and moves at a great height all the way to the poles, where it descends. This circulation, in fact a convection cell, is now known as a Hadley cell, but Hadley's idea of it has been much modified.

George Hadley (1685–1768) was an English meteorologist, although he trained originally as a lawyer. His brother John invented the forerunner of the sextant, the instrument navigators use to calculate latitude by measuring the angle between the Sun or a star and the horizontal.

1738

Daniel Bernoulli demonstrated that as the velocity of a flowing liquid or gas increases, its pressure decreases. This is known as "Bernoulli's principle" and has many applications, the best known of which is the aerofoil design by which the wings of an aircraft, or rotors of a helicopter, generate a strong lifting force. Bernoulli also tried to explain the relationship between the pressure, temperature, and volume of gases. These had already been observed, but no one knew why gases behave as they do.

Daniel Bernoulli (1700–82) was born at Groningen, in the Netherlands, into a Swiss family of mathematicians and physicists. His uncle Jacob (sometimes called Jacques) was almost as great a mathematician as Newton and Leibniz, and his father, Johann (or

Figure 5: *Daniel Bernoulli* (The Free Library of Philadelphia)

Jean) was only slightly less accomplished. Daniel had two brothers, two nephews, a cousin, and several other relations, all of whom were distinguished mathematicians or physicists. It was a very remarkable family. Daniel qualified in Switzerland as a doctor of medicine in 1724 and also trained as a mathematician. In 1725 he was appointed professor of mathematics in St Petersburg, Russia. He returned to Switzerland in 1733 as professor of anatomy and botany at the University of Basel, where he remained until his death.

1742

Celsius proposed the temperature scale that carries his name, although it is still sometimes referred to as the "centigrade" scale, the name that was officially abandoned internationally in 1948. As Celsius first described it, the scale had the boiling point of water as 0° and the freezing point as 100°, but this was soon reversed. Although the Fahrenheit scale is still used in some English-speaking countries, scientists throughout the world use the Celsius scale or Kelvin scale (in which 1 K = 1°C).

Anders Celsius (1701–44) was a Swedish astronomer. In 1730 he became professor of astronomy at Uppsala University, in the city where he was born and died.

1752

Benjamin Franklin performed his famous experiment, with a kite, which he used to prove that storm clouds carry electric charge and lightning is an electric spark. Attached to his kite was a pointed wire to which he had fastened a length of wet silken thread with a metal key tied at the bottom. During a thunderstorm he flew the kite toward the base of the cloud and held his hand close to the key. Sparks flashing from the key to his hand showed that a current had flowed from the cloud and down the silken thread. Franklin was lucky. The next two people who tried repeating his experiment were killed by it. The experiment led Franklin to suggest that if pointed metal rods were fixed on the highest points of buildings, with wires from the rods to the ground, the electric charge in storm clouds would be carried away safely and the buildings protected against damage by lightning. By 1782, about 400 lightning conductors had been fitted to buildings in Philadelphia alone. Franklin also calculated the tracks followed by storms crossing North America and was the first to study the warm-water Atlantic current now known as the Gulf Stream.

Benjamin Franklin (1706–90) was born in Boston. In the United States he is best known as one of the founding fathers of the nation, but he also became famous in Europe as a scientist and political philosopher. He invented a wood-burning stove and improved bifocal glasses, but his greatest work was in his studies of electricity. In 1776, Franklin was one of the group of Americans sent to

represent the emerging United States of America in France, where he became very famous and highly popular. He held many important positions and received many honors. Franklin died in Philadelphia.

1761

Joseph Black performed experiments which confirmed that when ice melts, it absorbs heat with no increase in its own temperature. He never published his finding, but from 1761 he included it in his lectures at the University of Glasgow, Scotland, and in 1762 he described it to a Glasgow literary society. In 1764 he and his assistant William Irvine (1743–87) measured the amount of heat absorbed and released when liquid water evaporates and water vapor condenses. The heat absorbed and released when water changes from one phase to another (ice to liquid to vapor) he called "latent heat." Latent heat plays a very important part in the formation of clouds and the development of thunderstorm clouds (cumulonimbus).

Joseph Black (1728–99) was a Scottish physicist, chemist, and medical doctor. He was born in Bordeaux, France, the son of a wine merchant. He was sent to be educated first in Belfast, then at the University of Glasgow, where he studied medicine and natural sciences. In 1751 he transferred to the University of Edinburgh to complete his medical studies. The thesis he wrote in 1754 for his doctor's degree described how heating magnesium carbonate (known then as "magnesia alba") releases a gas, distinct from air, which he weighed. Expanding on this work, in 1756 Black publish-

Figure 7: *Joseph Black*   (The Free Library of Philadelphia)

ed *Experiments Upon Magnesia Alba, Quicklime, and Some Other Alcaline Substances*, proving that when carbonates are heated, they lose this gas and become more alkaline, and when they absorb the gas, they become less alkaline. He called the gas "fixed air."

The gas had been discovered early in the 17th century by the Belgian chemist and alchemist Jan Baptista van Helmont (1577–1644), who called it "gas sylvestre" (because he obtained it by burning charcoal; the Latin *sylvestris* means "of the woods"). It is known today as carbon dioxide. (Helmont also coined the word *gas*; he called vapors by the Greek word *chaos*, but spelled it the way it sounded in Flemish, as "gas.")

In 1756, Black was appointed a lecturer in chemistry at Glasgow and also professor of anatomy, although he exchanged this post for the professorship of medicine and practiced as a physician. In 1766 he became professor of chemistry at the University of Edinburgh, the city where he died.

1783

Horace Bénédict de Saussure invented the hair hygrometer, still used today. This instrument exploits the fact that a human hair, with all the natural oils removed, increases in length by about 2.5 percent as the relative humidity increases from 0 to 100 percent. The hair hygrometer is small (the hair can be wound around a core to save space); with a simple system of gears, its reading can be displayed by a needle on a dial, and it can be manufactured cheaply.

Horace Bénédict de Saussure (1740–99) was a Swiss physicist. From 1762 to 1786 he taught at the University of Geneva, the city in which he was born and died. He was a keen botanist and Alpine explorer and wrote an account of his travels in the Alps.

1803

Luke Howard proposed a scheme for classifying clouds into four main and several secondary types and giving a Latin name to each. In his classification the main types were called stratus (Latin *stratum*, meaning "layer"), cumulus (the Latin word for "heap"), cirrus (the Latin word for "hair"), and nimbus (the Latin word for "cloud", but used here to describe the type of cloud that usually brings rain or snow). His classification, contained in an article called "On the Modification of Clouds," is the basis of the one still used today.

Luke Howard (1772–1864) was an amateur meteorologist who lived in London, England, and earned his living as an apothecary (druggist).

1806

Admiral Francis Beaufort proposed a scale for wind forces, allotting them values from 0 to 12 and specifying the amount of sail warships should carry under winds of each force. In 1838 the British

Figure 8: *Sir Francis Beaufort* (The Free Library of Philadelphia)

Admiralty adopted the scale officially, and in 1874 it was adopted by the International Meteorological Committee. In 1955 the U.S. Weather Bureau added forces 13 to 17 to describe hurricane-force winds. The Beaufort scale is still used, especially at sea.

Sir Francis Beaufort (1774–1857) was born in Ireland and was on active service in the Royal Navy for more than 20 years. In 1829 he was appointed hydrographer to the navy.

1820

John Daniell, an English chemist and meteorologist, invented the dew point hygrometer described in an article in the *Quarterly Journal of Science*

John Frederic Daniell (1790–1845) was born and died in London (while attending a meeting of the council of the Royal Society). In 1830 he was elected to the Royal Society, and the following year he became the first professor of chemistry at Kings College London, which had just been founded. He described his studies of meteorology in *Meteorological Essays*, published in 1823. Also see 1824 and 1830 below.

1824

John Daniell (see 1820) published "Essay on Artificial Climate Considered in Its Applications to Horticulture," in which he showed that it is important to maintain a moist atmosphere in hothouses growing tropical plants.

1827

J.-B. Fourier published a paper in which he compared the influence of the chemical composition of the atmosphere on its temperature to the effect of heating a bowl lined with black cork and covered with glass. This may have been the first account of what is now called the "greenhouse effect."

Jean-Baptiste Joseph Fourier (1768–1830) was a French mathematician, most widely known for his development of partial differential equations as a powerful scientific tool. He was born at Auxerre, in central France, the son of a tailor who died when Fourier was eight. He wanted to be a soldier, and after a mixed early education studied at the École Normale in Paris, later joining its staff and also the staff of the École Polytechnique. Fourier accompanied Napoleon on his campaign in Egypt in 1798. Napoleon rewarded Fourier's mathematical discoveries by making him a baron. Following the fall of Napoleon and the restoration of the Bourbons, he received more honors and in 1822 became one of the secretaries of the Academy of Sciences. He died in Paris from a disease contracted while he was in Egypt.

1835

Gaspard de Coriolis discovered that anything (not just air and water) moving over the surface of the Earth and not attached to it will be deflected by inertia acting at right angles to its direction of movement. This came to be called the "Coriolis force" (abbreviated as CorF), but no force is involved in the mechanical sense; today it is usually known as the "Coriolis effect" (still abbreviated as CorF).

The Coriolis effect causes air in the northern hemisphere to flow in a clockwise direction around areas of high pressure and counterclockwise around areas of low pressure (these directions are reversed in the southern hemisphere). It also affects ocean currents, causing them to follow curved paths and producing the Ekman effect, discovered in 1905 by the Swedish scientist Vagn Walfrid Ekman (see 1905 below).

Gaspard Gustave de Coriolis (1792–1843) was born and died in Paris. He became an engineer and mathematician and taught at the École Polytechnique, first as an assistant professor and from 1838 as director of studies. His health was poor, but he made many important discoveries in mechanics.

1840

Louis Agassiz published *Études sur les glaciers*, describing the results of studies he made in 1836 and 1837. He observed that mountain glaciers are not static, but flowing. From this, and the location of boulders (called "erratics") scattered far from the rock formations to which they belonged, he concluded that in the past, glaciers had extended much farther and had pushed the erratics to where they were later found. This led him to conclude that at one time, in the geologically recent past, the whole of Switzerland and much of northern Europe had lain beneath a single ice sheet, much like the one still existing in Greenland, and there had been a "Great Ice Age." Later Agassiz found confirmatory evidence in Scotland and also in many parts of North America, indicating that the Ice Age had affected all of the northern hemisphere. This was the first clear indication that climates in the relatively recent past had been very different from those of the present and, therefore, that climates can and do change over time.

Jean Louis Rodolphe Agassiz (1807–73) was born at Motier, on the shores of Lake Morat, Switzerland, the son of the local pastor. His mother taught him about natural history, and he developed a love of plants and animals. He was educated at Bienne and Lausanne, then at the Universities of Zürich, Heidelberg, and Munich. He obtained a doctorate of philosophy at Erlangen and a doctorate of medicine at Munich. In 1826 he was asked to complete the classification of fish specimens from Brazil (the scientist who started the work had died) and Agassiz became a leading authority on European fishes, an interest which led him to the study of fossil fishes, on which he became a world authority. In 1832 he went to study in Paris and then in Neuchâtel; with the help of Alexander von Humboldt and Georges Cuvier, two of the most eminent scientists of their day, he was appointed professor of natural history at the University of Neuchâtel, a post he held from 1832 to 1846.

In 1836, while on vacation in the Alps, Agassiz and his friends built a hut on the ice. After 12 years it had moved more than a mile (1.6 kilometer). He placed a line of stakes directly across the glacier, and found they also moved. In 1846 he was invited to lecture in the United States, at the Lowell Institute, Boston, in Charleston, and in several other cities. The lectures were highly popular, and Agassiz stayed in America, in 1838 becoming professor of zoology at Harvard. He is said to have been possibly the greatest science teacher ever to have worked in the United States. He was devoted to his students, whom he treated as colleagues, always emphasizing the importance of studying nature at first hand rather than from books. His work on fossil fishes helped support Darwinism, but Agassiz himself was always unsympathetic to the Darwinian view, possibly because he misunderstood it.

Agassiz died at Cambridge, Massachusetts, and is buried at Mt. Auburn. Beside his grave is a boulder from a Swiss glacial moraine.

1842

Matthew Maury was appointed the first director of the U.S. Naval Observatory and Hydrographic Office. He organized the collection of information from merchant ships on winds and currents. From these he discovered the shape of storms. His work led to an international conference on meteorology held in Brussels in 1853.

Matthew Fontaine Maury (1806–73) was born in Spotsylvania County, Virginia. He became a naval officer, but an accident rendered him permanently lame and unfit for active service. During the Civil War Maury sided with the South. After the war he spent some time in Mexico and England. In 1868 he became professor of meteorology at the Virginia Military Institute. He died at Lexington, Virginia.

C. J. Doppler discovered the Doppler effect, which is now applied to the study of weather systems by means of Doppler radar. Doppler found that if a source emitting waves at a constant frequency moves toward an observer, the frequency of its emissions will be shortened, and if the source moves away from an observer, the wavelength will be increased. His idea was tested by the Dutch meteorologist C. H. D. Buys Ballot at Utrecht in 1845, using an open-topped railcar and two groups of musicians. One set of musicians sat on the railcar, all playing the same, constant note on trumpets while the car moved past the other group, composed of people chosen because they possessed perfect pitch, standing beside the track. The listeners reported that as the railcar approached, the pitch of the trumpet note rose (implying a reducing wavelength), and after it had passed, the pitch fell (implying an increasing wavelength). The Doppler effect applies to all radiation, not only sound. Light becomes bluer if its source is approaching (shortening wavelength) and redder if it is receding (lengthening wavelength). This application is widely used by astronomers measuring the motion of distant galaxies in relation to Earth.

Christian Johann Doppler (1803–53) was born in Salzburg, Austria, and educated there and in Vienna. At first he earned his living as a schoolteacher; when he was 32 he thought of emigrating to the United States, but was then offered a better teaching position in Prague. This led, in 1841, to his appointment as professor of mathematics at the State Technical Academy in Prague. The paper announcing his discovery in 1842 was called "Über das farbige Licht der Doppelsterne" ("About the colored light of double stars"). In 1850 he was appointed director of the Physical Institute and professor of experimental physics at the University of Vienna. He died in Venice, which in those days was ruled by Austria.

Figure 9: *Samuel Morse*
(The Free Library of
Philadelphia)

1844

The first telegraph line in the world was constructed between
Baltimore and Washington. Funded reluctantly by the U.S. Congress, it cost $30,000. Samuel Morse claimed to have invented it,
and he certainly devised the code used to transmit messages. The
first message it carried was "What hath God wrought?" Telegraph
networks were soon installed in many countries and made possible
the collection of weather observations obtained simultaneously in
widely scattered locations. Until then, information could be transmitted no faster than the rider of a galloping horse could carry it.

Samuel Finlay Breese Morse (1791–1872) was born in Char-
lestown, Massachusetts. In his early years he was an artist, studying
art in England and specializing in portraiture after he returned to
the United States. He became well known, but not wealthy. To
transmit telegraph messages he devised the binary code of dots and

dashes that bears his name. Following the success of the telegraph system, Morse became involved in long and bitter disputes over his claim to have invented it, but eventually he was awarded the patents. He died in New York City.

1846

Joseph Henry was elected the first secretary of the Smithsonian Institution. He used the resources of the Smithsonian to obtain weather reports from all over the United States. The system Henry devised provided the basis on which the U.S. Weather Bureau was established.

Joseph Henry (1797–1878) was the most eminent American physicist of his day. He invented a telegraph in 1835, before Morse, but failed to patent it, and he helped Morse freely. He was born in Albany, New York, and died in Washington, D.C.

Figure 10: *Joseph Henry* (The Free Library of Philadelphia)

1851
At the Great Exhibition in London, the first weather map was published, showing readings taken simultaneously by instruments at many locations and collated at a central point. It was called a "synoptic" map, from Greek *syn*, "together," and *opsis*, "seeing." Weather maps covering a wide area are still known as synoptic charts.

1855
Urbain Leverrier began supervising the installation of a network to collect meteorological data from astronomical observatories throughout Europe.

Urbain Jean Joseph Leverrier (1811–77) was born in Saint-Lô and died in Paris. He started his working life as a chemist, but almost accidentally became an astronomer, teaching at the École Polytechnique in Paris. He made many important astronomical discoveries.

1856
William Ferrel proposed that low-latitude winds blowing toward the equator are deflected not by the rotation of the Earth, but by the tendency of moving air to rotate about its own axis. Once this begins, the rotation will be maintained to conserve angular momentum. In 1857 Ferrel also proposed on theoretical grounds what is now known as Buys Ballot's law. This was some months before Buys Ballot discovered it. William Ferrel (1817–91) was an American meteorologist.

1857
Buys Ballot discovered that in the northern hemisphere, winds circulate counterclockwise around areas of low pressure and clockwise around areas of high pressure. Therefore, if you stand with your back to the wind, the area of atmospheric low pressure will lie to your left and the area of high pressure to your right. In the southern hemisphere, these directions are reversed. This is known as Buys Ballot's law. Buys Ballot discovered it by observation, but it had been calculated theoretically several months earlier by the American meteorologist William Ferrel. Buys Ballot acknowledged this, but nevertheless the law continues to bear his name.

Christoph Hendrick Didericus Buys Ballot (1817–90) was a Dutch meteorologist. He founded the Royal Netherlands Meteorological Institute and, from 1854 until he died, was its director.

1861
In England, the Meteorological Department of the Board of Trade issued the first storm warnings for coastal areas on February 6 and for shipping on July 31.

John Tyndall published a paper in the *Philosophical Magazine and Journal of Science* showing that atmospheric gases absorb heat

Figure 11: *John
Tyndall* (The Free
Library of Philadelphia)

and, therefore, that the chemical composition of the air affects
climate. This was one of the first references to what is now called
the "greenhouse effect."

John Tyndall (1820–93) was an Irish physicist. He left school to
work as a surveyor and then as a civil engineer. He read widely
and attended scientific lectures whenever he could and eventually
became a student at the University of Marburg, Germany. He was
elected to the Royal Society of London in 1852, and in 1854 was
appointed professor of natural philosophy at the Royal Institution,
later becoming director of the institution. His most important work
concerned the conduction of heat and the scattering of light by
small particles. He found that this scattering makes it possible to
read in shadow, which would be impossible on the Moon, where

light is not scattered, and it allowed him to explain the blue color of the sky. Tyndall was a great popularizer of science, writing books and lecturing in language that nonscientists could understand. In 1872 and 1873 he made a lecture tour of the United States, donating the proceeds to a trust for the support of American science. He also studied glaciers and was one of the first people to climb the Matterhorn. He died in Hindhead, Surrey.

1863

The first network of meteorological stations linked by telegraph to a central point opened in France.

Francis Galton, in his book *Meteorographica*, introduced the word "anticyclone" and devised a method for mapping weather systems that is the basis of the one still used today. To do this, he sent a questionnaire to weather stations in various parts of Europe asking for the results of measurements made on a specified date, and plotted these on a map.

Sir Francis Galton (1822–1911) was an English geographer, anthropologist, and statistician, but with wide interests and a passion for scientific investigation. He was born at Sparkbrook, Birmingham, England, into a Quaker family and educated at Birmingham General Hospital, King's College London, Trinity College at Cambridge University, and St George's Hospital, London. In 1844, the year he graduated from Cambridge, his father died and left him a fortune large enough to finance his work. Galton traveled through the Balkans and Near East, then explored parts of southwest Africa; he advanced the science of meteorology, but is best known for his anthropological studies, inspired by the publication in 1859 of *On the Origin of Species* by Charles Darwin, a cousin of Galton. Galton sought to discover the extent to which human characteristics are inherited or produced by the conditions under which children are brought up. To help make sense of the data he collected, in 1888 Galton invented a way to calculate the "correlation coefficient," an important statistical technique. Galton invented "eugenics," the idea that superior humans can be produced by selective breeding, and was the person mainly responsible for introducing fingerprinting for the identification of criminals. He was knighted in 1909 and died at Haslemere, Surrey.

1869

From September 1, the first weather bulletins were issued on a daily basis from the Cincinatti Observatory at the instigation of the observatory director, Cleveland Abbe. The service proved very popular and led to the establishment of a national bureau, headed by an army general and, from 1871, with Abbe as scientific assistant.

Cleveland Abbe (1838–1916) became known as the "father of the Weather Bureau." He taught meteorology at Johns Hopkins

University and conducted extensive research. He was born in New York and died in Chevy Chase, Maryland.

1871

As scientific assistant at the Weather Bureau, Cleveland Abbe began issuing three-day weather forecasts.

1874

The International Meteorological Congress was founded. In 1896 it published its standard classification of cloud types in the *International Cloud Atlas*, which has been revised and updated several times. This work is now published by the World Meteorological Organization (WMO) of the United Nations.

1875

The Times of London published the first weather map to appear in a newspaper. It had been drawn by Francis Galton.

1878

In England, the Meteorological Office published its first *Weekly Weather Report* on February 11.

1884

S. P. Langley published a paper on the climatic effect of the absorption of heat by atmospheric gases, an early reference to what is now known as the "greenhouse effect." Langley had also measured the spectrum of light reflected by the Moon at different seasons and with the Moon at different heights above the horizon. This allowed Svante Arrhenius (see 1896) to calculate how much heat carbon dioxide and water vapor absorb.

Samuel Pierpont Langley (1834–1906) was an American astronomer, born at Roxbury, Massachusetts. He largely educated himself, but did so well enough to be made an assistant at Harvard University and eventually to receive professorships at various establishments. In 1887 he became secretary of the Smithsonian Institution. In 1881, Langley invented a bolometer, an instrument for measuring very small quantities of heat very accurately, and used to measure the amount of solar radiation. He was also enthusiastic about aviation and in 1896 built a model aircraft powered by steam, which flew about three-quarters of a mile (1.2 kilometers). The government gave him a total of $50,000 to build a full-size model, but the materials he used were not strong enough to survive the stresses, and the three trials he made between 1897 and 1903 all failed. This led the *New York Times* to attack him in an editorial for wasting public money; the editorial predicted that humans would not fly for a thousand years. Nine days later the Wright brothers did precisely that. Langley died in Aiken, South Carolina.

Figure 12: *Samuel Langley*
(The Free Library of
Philadelphia)

1891

The weather bureau where Cleveland Abbe worked as scientific
assistant became the U.S. Weather Bureau, with Abbe as the
meteorologist in charge.

1893

Edward Maunder discovered that the coldest part of the Little Ice
Age, between 1645 and 1715, coincided with a period of very low
sunspot activity, not a single sunspot having been reported for 32
years. This is now known as the "Maunder Minimum," and earlier
Maunder Minima have also been found to coincide with periods

when the climate cooled appreciably, though not to the extent of a full ice age.

Edward Walter Maunder (1851–1928) was a British astronomer. He made his discovery by searching through old records while working as superintendent of the solar division of the Royal Greenwich Observatory, London.

1896

In April, Svante Arrhenius published an article in the *Philosophical Magazine and Journal of Science* in which he linked changes in the atmospheric concentration of carbon dioxide with climate. This is now called the "greenhouse effect." To do this, he performed somewhere between 10,000 and 100,000 calculations (with no help from a calculator or computer!) and predicted that a doubling of the carbon dioxide concentration would cause the average global temperature to rise by 9–11°F (5–6°C) (the present estimate is 2.7–8.1°F) (1.5–4.5°C). He thought changes in carbon dioxide were due mainly to changes in the frequency, magnitude, and type of volcanic eruptions, and that these might cause ice ages to start and end. He estimated it would take 3,000 years to double atmospheric CO_2 by burning fossil fuels. Arrhenius thought such a global warming would be beneficial, because it would increase crop yields and raise standards of living.

Arrhenius was not the first scientist to suspect a link between climate and the chemical composition of the atmosphere (see 1827, 1861, and 1884), but he was the first to calculate the change that would result from an increase in atmospheric carbon dioxide.

Svante August Arrhenius (1859–1927) was a Swedish physical chemist. He was born in Wijk, near Uppsala, and by the age of three had taught himself to read. He studied at the University of Uppsala, then moved to the University of Stockholm to work for a higher degree. His most famous work was on the conduction of an electric current through certain solutions, called electrolytes. For this he was awarded the 1903 Nobel Prize in Chemistry. His interests were very wide, and in *Worlds in the Making*, a book he published in 1908, he suggested life on Earth developed from living spores that arrived from space (a theory known as "panspermia"). In 1895 Arrhenius was appointed a professor at the University of Stockholm, and in 1905 he became director of the Nobel Institute for Physical Chemistry, a post he held until shortly before his death. He died in Stockholm.

1902

L. P Teisserenc de Bort discovered the stratosphere, using balloons carrying instruments which revealed that the atmosphere has two layers. In the lower layer temperature decreases with height. He called this the "troposphere" (from the Greek *tropos*, meaning "turning"). Above about 7 miles (11.2 kilometers), he found temperature remained constant with height. He called this layer the "stratosphere" (from the Latin *sternere*, meaning to "strew" or "spread in layers"), and he called the boundary between these layers the "tropopause." He suggested that in the stratosphere gases might form layers, with oxygen at the bottom, nitrogen (which is lighter) above it, helium above that, and hydrogen at the top. In fact, gases are not arranged in layers in the stratosphere.

Léon Philippe Teisserenc de Bort (1855–1913) was born in Paris, the son of an engineer. In 1880 he started work in the meteorological department of the Central Bureau of Meteorology, in Paris, and in 1892 became chief meteorologist of the bureau. He resigned in 1896 to found a private meteorological observatory in Trappes, near Versailles, which is where he conducted his experiments. He died in Cannes.

Vilhelm Bjerknes published *Weather Forecasting as a Problem in Mechanics and Physics*. This was one of the first scientific studies of weather forecasting.

Vilhelm Frimann Koren Bjerknes (1862–1951) was a Norwegian meteorologist. He was born and died in Oslo. As a student, he helped his father, who was professor of mathematics at Christiania University (now the University of Oslo). He himself held professorships at Stockholm and Leipzig before returning to Norway in 1917 to found the Bergen Geophysical Institute. During World War I, Bjerknes established a series of weather stations throughout Norway. Information from these allowed Bjerknes and his colleagues, who included his son, Jakob, and Tor Harold Percival Bergeron (1891–1977) to develop their theory of air masses bounded by fronts.

1905

By watching the movement of floating sea ice, V. W. Ekman discovered that it, and any wind-driven sea current, moves at 45° to the right of the wind direction in the northern hemisphere (and to the left in the southern). This motion is due to the combined effects of the wind direction and strength, the Coriolis effect, and friction between different layers of water. With increasing depth, currents are slowed and deflected farther to the right in the northern hemisphere (and to the left in the southern hemisphere), so there is a depth at which the current flows in the opposite direction to the surface current. This "Ekman depth" varies, but averages about 165 feet (50 meters) over much of the ocean. Between the surface and this depth, the current direction describes a spiral, known as the "Ekman spiral." It was later discovered that wind direction also forms an Ekman spiral with increasing altitude.

Vagn Walfrid Ekman (1874–1954) was a Swedish oceanographer. He spent several years working at the International Laboratory for Oceanographic Research in Oslo, Norway, and returned to Sweden in 1908. In 1910 he was appointed professor of mathematical physics at Lund University. His discovery of the Ekman spiral arose from an observation made in the 1890s by the Norwegian arctic explorer Fridtjof Nansen, that drifting sea ice moved in a direction about 45° to the right of the wind direction.

1913

Charles Fabry (1867–1945), a French physicist, discovered the ozone layer of the stratosphere.

1918

W. P. Köppen published a system for classifying climates according to the vegetation typical of them. He divided climates into six groups, broadly based on temperature and the seasonality of precipitation. A winter temperature of 64°F (17.7°C) is critical for certain tropical plants, for example, a summer temperature of 50°F (10°C) is necessary for trees, and a temperature of 27°F (-2.8°C) indicates snow at some time every winter. His climate groups included the following: tropical rainy; arid; warm, temperate, rainy; rainy; tundra; and permanent frost and icecaps. He later modified his system, completing the modifications in 1936. The Köppen classification is still widely used.

Wladimir Peter Köppen (1846–1940) was born of German parents in St. Petersburg, Russia. He studied at the Universities of Heidelberg and Leipzig and from 1872–73 worked in the Russian meteorological service. In 1875 he moved to Hamburg, Germany, where he headed a new division of the Deutsche Seewarte formed to issue weather forecasts for the land and sea areas of northern Germany. He was able to devote himself entirely to research from 1879. He died in Graz, Austria.

1921

Vilhelm Bjerknes published *On the dynamics of the circular vortex with applications to the atmosphere and to the atmospheric vortex and wave motion*, in which he established that the lower atmosphere is composed of air masses. These are distinguished from one another by temperature, pressure, and humidity. (See 1904.)

1922

L. F. Richardson published *Weather Prediction by Numerical Process*, in which he described a numerical method for forecasting the weather. His method was based on the idea that a rise or fall in surface atmospheric pressure reflects a convergence or divergence of air throughout the column of air extending above the surface all the way to the tropopause. By applying the laws of physics to the observed situation, Richardson believed it possible to calculate how the atmosphere would change for hours or days ahead. His system failed, partly for want of accurate measurements of conditions in the upper atmosphere and partly because the convergence and divergence of air varies at different heights, and these differences are much more important than overall convergence and divergence throughout the column. The method was also very laborious, necessitating many separate calculations that in those days had to be made without the help of calculators and computers, although Richardson invented a cylindrical, hand-held calculator to help.

A variant of his "numerical forecasting" was introduced in the 1950s, however, and is still used. The first routine forecasts calculated in this way were made in 1955 in the United States and 1965 in Britain. Modern numerical forecasting uses data obtained from measurements made at many different heights and calculates changes, and their effects on one another, at each level; it also allows for many more influences on the weather than was possible earlier. So many calculations are required that numerical forecasting became practicable only with the introduction of fast supercomputers. Without them, it took so long to prepare the forecast that the predicted weather would pass before the forecast could be issued. Lewis Fry Richardson (1881–1953) was a British mathematician and meteorologist, born at Newcastle upon Tyne into a Quaker family and educated in York and at Cambridge University. Many of the ways he found of applying mathematics to complex problems were imaginative and years ahead of their time. He even attempted to use mathematics to discover the causes of war. He died at Kilmun, Argyll, Scotland.

1923

Sir Gilbert Walker had described a high-level movement of air close to the equator that flows from west to east, counterbalancing the trade winds blowing near the surface. This upper-air flow is called the "Walker circulation," and in 1923 Walker published a description of periodic changes in the distribution of tropical air pressure that are associated with it. He called these the "Southern Oscillation." By the 1970s the Southern Oscillation was recognized as contributing to the appearance of El Niño, a current in the tropical South Pacific linked to major climatic events. Today both are considered a single phenomenon, called an El Niño–Southern Oscillation (ENSO) event.

Walker was a British meteorologist who was appointed director of the Indian meteorological service in 1904. There had been severe famines in 1877 and 1899 due to the failure of the Indian monsoon, and Walker took a particular interest in the causes of the Asian monsoons and their occasional failure. It was his study of these that led him to his discovery of the Southern Oscillation.

1930

Milutin Milankovich proposed that three variations in the movement of the Earth in relation to the Sun can coincide to trigger the commencement and ending of ice ages. His theory is now widely accepted by climatologists.

Earth follows an elliptical path in its solar orbit. Over a cycle of about 100,000 years, the ellipse lengthens, then shortens again, altering the amount of solar radiation received at the surface.

Over a cycle of about 40,000 years, the Earth's axis of rotation moves in a small circle, like a wobbling gyroscope, altering the angle between the axis and the radiation coming from the Sun. Over a cycle of about 21,000 years, the date at which the Earth is closest to the Sun (perihelion) in its orbit, it moves through a complete year; at present perihelion is reached in January and 10,000 years from now it will be reached in July. This alters the intensity of solar radiation received at the surface in winter and summer.

Figure 13: *Wilson Bentley* (The Free Library of Philadelphia)

Milutin Milankovich (1879–1958) was a Serbian climatologist. He was educated in Vienna, but moved to the University of Belgrade in 1904 and remained there for the rest of his life.

1931

Wilson W. Bentley published *Snow Crystals*, a book containing more than 2,000 of the more than 5,000 photomicrographs (photographs taken through a microscope) he had taken of snowflakes. His photographs were of such high quality that they aroused scientific interest in the subject and contributed to the development of an international system for classifying types of snowflakes.

Wilson W. Bentley (1865–1931) was an American farmer and meteorologist who lived in Jericho, northern Vermont, where a lot of snow falls every winter. He was entranced by the beauty of snowflakes and spent many hours patiently photographing the many varieties of them.

C. W. Thornthwaite published a system for classifying climates. He divided climates into groups according to the types of natural vegetation typical of them. This is determined by what Thornthwaite called "precipitation effectiveness," calculated by dividing the total monthly precipitation (P) by the total monthly evaporation (E). The 12 monthly values are then added to produce a P/E index, from which five "humidity provinces" are defined. A P/E index of more than 127 (called "wet") indicates rain forest; 64–127 (humid) is forest; 32–63 (subhumid) is grassland; 16–31 (semi-arid) is steppe; and less than 16 (arid) is desert. In 1948, Thornthwaite revised his system to incorporate a "moisture index," which relates the water needed by plants to the available precipitation by calculating an index of potential evapotranspiration (PE). He also included an index of "thermal efficiency," calculated from monthly temperatures, with 0 indicating a frost climate and 127 a tropical climate.

Charles Warren Thornthwaite (1889–1963) was an American climatologist who taught at the Universities of Oklahoma (1927–34) and Maryland (1940–46), and at Johns Hopkins University (1946–55) before becoming director of the Laboratory of Climatology at Centerton, New Jersey, and professor of climatology at Drexel Institute of Technology, Philadelphia. Thornthwaite was president of the Section of Meteorology of the American Geophysical Union from 1941 to 1944, and in 1951 he was elected president of the Commission for Climatology of the World Meteorological Organization.

1940

Carl-Gustav Rossby discovered large undulations in the westerly winds of the upper atmosphere. These are now known as "Rossby waves," and are also known to occur in the oceans. Rossby showed that the upper winds have a powerful effect on the weather. When they blow strongly, active frontal systems develop beneath them, producing stormy conditions. When they are weak, cold air is able to move south.

Carl-Gustav Arvid Rossby (1898–1957) was one of the most eminent meteorologists of this century. He made major contributions to our understanding of the behavior of air masses and air movements. After 1954 he initiated and led worldwide studies in atmospheric chemistry. Rossby was born and educated in Stockholm, Sweden, but moved to the United States in 1925. He worked in Washington, D.C., then became professor of meteorology at the Massachusetts Institute of Technology. He also held a professorship at the University of Chicago. He returned to Sweden in 1948 and founded the Institute of Meteorology, in association with the University of Stockholm.

1946

Vincent Schaefer discovered that pellets of dry ice (solid carbon dioxide) at about -9.5°F (-23°C) injected into moist air caused water vapor to sublime into ice crystals. Like many discoveries, this one was made by accident. Schaefer and his colleague, Nobel laureate Irving Langmuir (1881–1957), were working at the General Electric Research Laboratory at Schenectady, investigating the crystallization of ice from moist air, a topic of some importance because of problems caused by the icing of airplane wings. They used a refrigerated box in which they attempted to make crystals form on various types of dust particles. There was a spell of very hot weather in July 1946, and keeping the temperature in the box low enough for the experiments became difficult. Schaefer tried chilling the air in the box by dropping dry ice pellets into it, and within moments a miniature snowstorm filled the box.

Realizing the importance of his discovery, on November 13, 1946, Schaefer tried a bigger experiment. He was flown above a layer of cloud over Pittsfield, Massachusetts, dropped 6 pounds (2.7 kilograms) of pellets into the cloud, and triggered a snowstorm. This success led to many attempts at modifying or controlling the weather and was the start of the science of experimental meteorology.

Vincent Joseph Schaefer (1906–93) was born and died in Schenectady, New York. He dropped out of school at 16 and went to work in the machine shop of General Electric, but later returned to his studies and graduated in 1928 from the Davey Institute of Tree Surgery, and for a time worked as a tree surgeon. Throughout his life he had a great love of the open air, but, although he enjoyed working with trees, financial pressures forced him to return to General Electric. Langmuir made Schaefer an assistant in 1933, and they worked together throughout World War II, inventing several devices that proved useful. Schaefer left General Electric in 1954 and, from then until 1958, was research director at the Munitalp Foundation, after which he devoted his time to research and education. He joined the faculty of the State University of New York in 1959; Schaefer became a founder of the Atmospheric Sciences Research Center in 1960, and was its director from 1966 to 1976.

1949
Radar was first used to obtain meteorological data as part of the U.S. Thunderstorm Project.

1951
An international system was adopted for the classification of types of snowflakes. It divides snowflakes into 10 principal types, each of which can be further divided into subtypes.

1954
Ukichiro Nakaya, of the University of Hokkaido, Japan, published *Snow Crystals*, the classic work on the form and structure of snowflakes.

1959
The U.S. Weather Bureau began publishing a temperature–humidity index (THI) as an indication of how comfortable or uncomfortable people will find a hot day.

1960
The first weather satellite, Tiros 1, was launched by the United States on April 1.

1964
The weather satellite Nimbus 1 was launched on August 24. This was the first satellite to produce high-quality photographs taken at night.

1966
The first meteorological satellite to be placed in geostationary orbit was launched over the Pacific on December 6.

1971
T. Theodore Fujita and Allen Pearson devised a standard six-point scale for reporting tornado intensity. The Fujita Tornado Intensity Scale relates wind speed to the damage caused. (See Wind Measurement and Cloud Classification, p. 128.)

Tetsuya Theodore Fujita (he adopted the name Theodore in 1968) is professor of meteorology at the University of Chicago. Allen Pearson was formerly the chief tornado forecaster for the U.S. National Weather Service.

1974

The first of the Geostationary Operational Environmental Satellites (GOES) was launched.

F. Sherwood Rowland (now at the University of California at Irvine) and Mario Molina (now at the Massachusetts Institute of Technology) proposed that chlorofluorocarbon (CFC) compounds used as aerosol propellants, working fluids in refrigerators, freezers, and air conditioners, and in making foam plastics, may survive in the atmosphere long enough for significant amounts to penetrate the stratosphere. There they might engage in a chain of chemical reactions leading to the depletion of ozone. For this work, they shared with Paul Crutzen (now at the Max Planck Institute for Chemistry, at Mainz, Germany) the 1995 Nobel Prize for Chemistry.

1977

Meteosat, the first European meteorological satellite to be placed in geostationary orbit, was launched by the United States on November 23. It remained functional until 1985.

1989

The European Space Agency launched the French-built Meteosat-4.

1992

The Topex–Poseidon satellite was launched, carrying instruments to measure very accurately the height of sea level. The satellite achieves this with two measurements: One uses a radar signal to measure the distance between the satellite and the ocean surface; the second measures the gravity field of the Earth and calculates from this what the distance from the satellite to the ocean surface would be if the ocean were still. By subtracting one measurement from the other, oceanographers can calculate the height of waves and track the movement of ocean currents. The Topex–Poseidon satellite also allows scientists to observe changes in the equatorial current that precede the onset of an ENSO. This means ENSO events can be predicted more quickly.

1995

On June 3, Joshua Wurman and Jerry M. Straka, scientists at the School of Meteorology at the University of Oklahoma, used new equipment to observe the internal structure of a tornado near Dimmitt, Texas, in more detail than had been possible previously. They used a pencil-beam Doppler radar mounted on a small truck. It transmitted a beam only 1.2° wide from a range of 1.2–3.7 miles, or 1.92–5.92 kilometers, (the range changing as the tornado moved) and revealed the structure of the tornado, the wall of debris surrounding the core, wind speeds, and the speed (more than 56 MPH or 90 KPH), of the central downdraft.

Experiments you can perform at home

Scientists learn about the world and the universe methodically. They begin by observing closely what happens. Then they try to find an explanation that fits their observations. This explanation is called a "hypothesis," and it must be tested to find out whether it is true. The aim is not to prove directly that the hypothesis is true, but to try to prove it false. If all attempts to prove it false fail, the hypothesis will come to be accepted. It will then be called a "theory."

Testing a hypothesis begins by making predictions. If the hypothesis is true, the scientist argues, under certain specified circumstances particular events will happen. The scientist sets up an experiment, therefore, in which he or she recreates those conditions and sees whether the predicted events occur. If they fail to do so, and whenever the experiment is repeated it continues to fail, the hypothesis must be rejected. As an explanation, it cannot be true. If the predicted events do occur, however, this does not necessarily support the hypothesis, because the events might have occurred by pure chance. This means that experiments must be repeated, often many times. Their results must be recorded carefully and then subjected to statistical tests, which calculate the likelihood that the results are due to chance. Only if the experiments pass these tests are their results considered to support the hypothesis.

Experimental results will be accepted by scientists generally only if the same results appear when the experiment is repeated by scientists other than the one(s) who first designed and performed it. When describing the results, the scientist must also explain how they were obtained in sufficient detail for anyone reading the account to repeat the experiment.

Science advances by exposing all the work scientists do to the toughest criticism other scientists can devise, but the criticism must always be of the work and the ideas underlying it, and never of a scientist personally. The way scientists describe their own work and comment on the work of others is very formal and not always easy for nonscientists to follow, but basically it is no different from the ordinary way we all evaluate ideas. When we are offered an explanation for something in the world around us, we ask ourselves whether the explanation sounds sensible, whether it would work in all cases, and whether there might be a simpler and better explanation. Scientists use experiments to check ideas, but that is the only real difference from the way we check ideas. Science is just a special kind of common sense.

The 30 experiments listed here can all be performed easily at home, and the equipment you will need can be assembled from ordinary household items. Although these experiments do not require high-tech equipment, they are not unscientific. Since science is just a special kind of common sense, it is defined by the approach you take to studying natural phenomena, not the cost of the equipment you use. Although they are not described in this way, all the experiments and demonstrations follow the same underlying method. They begin with a hypothesis and then test it. In the first experiment, for example, the hypothesis states that when air expands, it cools, and when it is compressed, it warms. If you perform the experiment, you will compress air and allow it to expand to see whether its temperature changes in the predicted way.

If you convert part of your home into a temporary laboratory, you must learn another very important lesson: safety. If you visit a professional laboratory or watch documentary television programs showing scientists at work, you will notice they all wear white coats. These are to protect their clothes; You will also notice scientists often wear what look like eyeglasses. There may be nothing wrong with their vision. These are safety goggles to protect their eyes from substances that may spurt into their faces. If scientists handle hazardous materials, they wear gloves.

Like the professionals, you must perform your experiments carefully and safely. Remember that hot items can cause burns and sharp edges can cut you.

Safety Guidelines

Always keep flames and hot liquids well away from your face and clothes.

Make sure flames are nowhere near materials they could ignite, such as papers.

Use pins, knives, and scissors carefully, never pointing them toward your face or body.

If you are unsure, ask an adult to help you.

Observe the safety rules and you will come to no harm.

The experiments

1. ADIABATIC WARMING AND COOLING

You will need: A bicycle wheel; a bicycle pump; a thermometer; sticky tape.

1.   Tape the thermometer to a spoke, with the bulb about 3 inches (7.62 centimeters) above the tire valve. Make sure you can read the thermometer easily.

2.   Wait a few minutes, then note the temperature. This is the air temperature.

3.   Remove the tire valve so all the air escapes from the tire. As it escapes, watch the thermometer. The temperature will fall because the pressurized air is expanding.

4.   When all the air has escaped, replace the valve.

5.   Remove the thermometer from the wheel spoke and tape it to the connector tube of the pump so that the bulb is pressed tightly against the tube, but do not cover the bulb with tape. Connect the tube to the valve.

6.   Wait a few minutes, then note the air temperature.

7.   Pump air into the tire as vigorously as you can until the tire is fully inflated. Watch the thermometer (or have a friend watch it for you). The temperature will rise as the air is compressed.

2. THE CORIOLIS EFFECT

You will need: A wooden board; a piece of cardboard about 12 inches (30.48 centimeters) square; a pair of compasses; ruler; scissors; hammer and nail; thumbtack; pencil.

1.   Cut a strip 1 inch (2.54 centimeters) wide from the edge of the card.

2.   Draw a circle with a radius of about 5 inches (12.7 centimeters) on the remaining card. Cut it out to make a disk.

3.   Place the disk on the wooden board, with the strip over it, and nail both to the board through the center of the disk so both can turn freely around the nail.

4.   Fix the other end of the strip to the board with the thumbtack. Make sure the disk turns easily beneath it.

5.   Using the strip as a ruler, draw a straight pencil line on the disk from the edge (the "equator") to the center (the "pole"), at the same time turning the disk steadily.

6.   The line you have drawn will be curved (the direction it curves depends on the direction you turned the disk), and the amount of curve will increase the farther it is from the edge.

3. FINDING THE PREVAILING WIND BY MAKING A WIND ROSE

(*See the drawing.*)

You will need: 12 square pieces of card, each side 2 feet (60 centimeters); a compass; a protractor; a ruler; a pen; a notebook.

1.   Draw lines to divide each of the square cards into four quarters. Mark the edges of the cards, where the dividing lines meet them, N, S, E, and W. Label the cards, one for each month of the year.

2.   Start on the first day of a month at a particular time. Find a high wind vane (e.g., on a church). This will indicate the wind at a height well clear of trees and buildings. Note the wind direction.

3.   From the center of the card (where the dividing lines intersect) for that month, use the protractor to mark the wind direction

Figure 14: *Experiment 3*

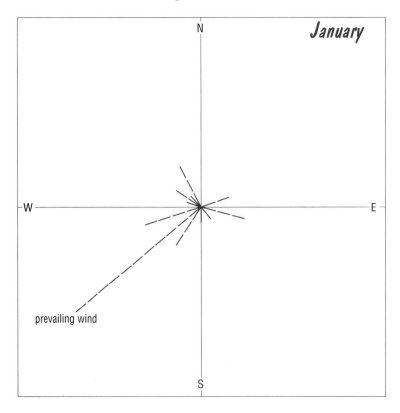

(remember, the direction the wind blows *from*). Draw a line .5 inch (1.3 centimeters) long from the center in that direction.

4.   Note the wind direction from the same wind vane at the same time every day.

5.   Each day, mark the wind direction with a .5-inch (1.3-centimeter) line on the card. If the direction is within about 10° of that for a previous day, so a line already exists, add the new measurement at the end of the existing line, making it .5 inch (1.3 centimeter) longer.

6.   After you have recorded the wind for the last day of the month, the longest line on the "wind rose" you have made indicates the prevailing wind during that month.

7.   Repeat this for a year with a separate wind rose for every month. By measuring the lengths of the lines and adding them together, you will be able to work out the prevailing wind during each month, each season, and for the year as a whole.

4. RELATIONSHIP BETWEEN THE TEMPERATURE AND VOLUME OF A MASS OF AIR

You will need: An ordinary balloon; a freezer.

1.   Fully inflate the balloon at room temperature and seal it firmly.

2.   Place the inflated balloon inside the freezer. Leave it there one hour.

3.   Remove the balloon and note how it has shrunk; it will look wrinkled. The amount (mass) of air it contains has not changed (it is still sealed), but that mass now occupies a smaller volume.

4.   Leave the shrunken balloon in the room at room temperature. Slowly it will return to its former size as the air inside it expands.

5. CONSERVATION OF ANGULAR MOMENTUM

You will need: A length (about 4 feet, or 1.2 meters) of strong string; a strong stick; a weight with a hole in it.

1.   Tie the weight firmly to one end of the string.

2.   Tie the other end of the string firmly to one end of the stick.

3.   Holding the stick, whirl the weighted string in a circle, around and around. Notice the speed at which it turns (you could ask a friend to time one complete turn).

4.   When it is whirling at a more or less constant speed, keep whirling it in the same way, but allow the string to wind itself around the stick, so the weight is drawn inward.

5. Notice what happens to the speed of the weight. It will accelerate as it approaches the stick, because its angular momentum is conserved.

6. THE BERNOULLI PRINCIPLE (1)

You will need: A penny.

1.   Place the coin on a smooth table, about .5 inch (1.3 centimeters) from the edge.

2.   Bend down so your mouth is level with the edge of the table.

3.   Blow hard across the top of the coin. It will jump because the air pressure above it suddenly falls, creating lift. (You can use this demonstration to try "flying" your penny into a saucer.)

7. THE BERNOULLI PRINCIPLE (2)

You will need: A spoon.

1.   Turn on a tap until there is a strong flow of cold water.

2.   Hold the spoon by the tip of its shaft with the bowl of the spoon pointing vertically down. Hold it just tightly enough to prevent the spoon dropping.

3.   Slowly move the spoon toward the water until the convex (bulging) side of the bowl touches the edge of the stream.

4.   Notice what happens. Instead of being pushed away by the flowing water, the spoon is pulled toward it, into the stream. This is because the pressure in the flowing water is lower than the air pressure outside it.

8. THE BERNOULLI PRINCIPLE (3)

You will need: Two sheets of typing paper, about 11.5×8 inches, or 29.2×20.3 centimeters (the exact size does not matter).

1.   Hold the two sheets of paper side by side, about 1 inch (2.54 centimeters) apart.

2.   Blow gently between them. Instead of the sheets moving away from one another, they are drawn together. This is because the pressure in moving air is lower than the pressure in the still air.

9. THE BERNOULLI PRINCIPLE (4)

You will need: One sheet of typing paper.

1.   Hold the paper so one edge is level with your lower lip and the rest of the sheet curves downward.

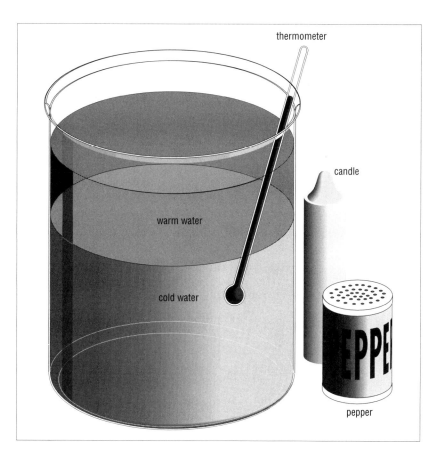

Figure 15: *Experiment 10*

2.   Blow gently over the sheet. It will rise upward because the pressure in moving air (your breath when you blow) is less than that in the still air beneath the paper.

10. AIR MASSES AND FRONTS

(*See the drawing.*)

You will need: A tall jar; a heatproof container; a thermometer; food coloring; ground black pepper; a candle.

1.   Half-fill the tall jar with cold water.

2.   Sprinkle ground pepper into the water and stir until the pepper is evenly distributed. Set the jar aside.

3.   Pour hot water into the heatproof container.

4.   Add enough food coloring to clearly tint the water, and stir until the color is even.

5.   Very gently, pour the hot water into the jar of cold water. It will form a clearly defined layer, and the two will mix only slightly.

6.   Check that they have not mixed by measuring the temperatures in the upper and lower layers.

7.   Light the candle and hold the jar above it. As the base warms the pepper will start to move upward, but it will not penetrate the overlying water, which is hotter than the water immediately below.

11. MEASURING DEW POINT TEMPERATURE

You will need: A large straight-sided, glass or metal container; a large bowl; crushed ice; a long-handled spoon; a thermometer.

1.   Make sure the outside of the container is completely dry.

2.   Fill the bowl with crushed ice (to make crushed ice, place ice cubes on a flat, solid surface; cover them with a cloth; and hit them sharply with a heavy object such as a hammer).

3.   Half-fill the container with cold water.

4.   Place the thermometer in the container of water.

5.   Add crushed ice to the container, a spoonful at a time. Stir the water after adding each spoonful of ice to mix it thoroughly.

6.   Watch the outside of the container closely. As soon as moisture starts to condense on it, note the temperature of the water. This is the dew point temperature.

12. CONVECTION CELL

(*See the drawing.*)

You will need: A large, tall-sided container (made from glass or clear plastic, if possible); cardboard; scissors; modeling clay; sticky tape; a candle; an incense stick; matches; a ruler; a pencil.

1.   Measure the width and height of the container.

2.   Cut the cardboard to fit the container, so it will make a partition, dividing the container into two sections.

3.   Cut two squares, each side about 2 inches (5.08 centimeters), from the corners of the cardboard, both on the same, short side.

4.   Fit the cardboard in the container, using sticky tape to hold it, so the container is divided into two compartments, with two square spaces on either side at the bottom.

5.   Use modeling clay to fix the candle in about the middle of one compartment.

6.   Light the candle.

7.   Light the incense stick.

8.   Hold the incense stick in the other compartment (the one without the candle), pointing down.

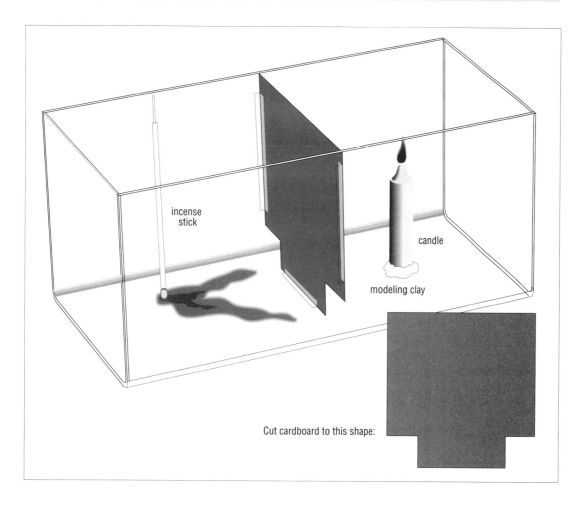

incense
stick

candle

modeling clay

Cut cardboard to this shape:

9. Watch how the smoke moves. It will rise up in the heated compartment and air will be drawn down in the unheated compartment, setting up a steady flow of air. This is a convection cell.

Figure 16: *Experiment 12*

13. SUNSHINE AND LATITUDE

(*See the drawing.*)

You will need: A spherical balloon; a flashlight; paper; adhesive tape. Perform this demonstration after dark.

1. Fold the paper to make a funnel. The narrow end should be about .5 inch (1.3 centimeters) in diameter, the wide end big enough to fit over the end of the flashlight. Tape the sides of the paper to keep the funnel in shape.

2. Tape the wide end of the funnel over the end of the flashlight.

3. Inflate the balloon.

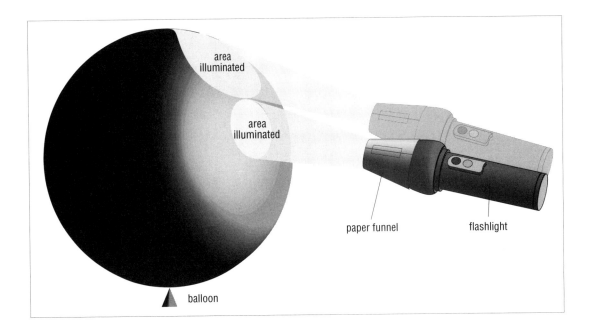

area
illuminated

area
illuminated

paper funnel   flashlight

balloon

Figure 17: *Experiment 13*

4. Make a small loop of tape, sticky side outside, and use it to fasten the balloon to the top of a table.

5. Turn off the lights and wait a few minutes for your eyes to grow accustomed to the dark.

6. Hold the flashlight level with the middle of the balloon (its "equator") and turn it on. Notice the size of the illuminated area on the balloon and how brightly it is lit.

7. Now tilt the flashlight so it shines on the "northern hemisphere" of the balloon. Notice the area it illuminates and how brightly it is lit.

14. THE EFFECT OF IMPURITIES ON THE FREEZING TEMPERATURE OF WATER

You will need: Two jars, both the same size; a spoon; a thermometer; salt; water; a freezer.

1. Fill both jars with cold water, to about three-quarters full.

2. Pour salt into one of the jars, stirring to dissolve it. If you add about 0.7 ounce (2 grams) of salt for every pint (.47 liter) of water, the result will be about as salty as sea water; you can speed up the demonstration by adding more.

3. Place both jars in the freezer.

4. Check them every 30 minutes until ice starts to form. Then note which jar freezes first.

5.   Each time you check the jars, measure the temperature in each, remembering to dry the thermometer and let it warm a little between jars. You will find the salt water takes longer to freeze than the water with no salt.

6.   When a layer of ice has formed on the salt water, remove some of the ice and taste its upper surface. You will find it tastes much less salty than the water. When salt water freezes, the ice consists almost entirely of fresh water; the salt is left behind.

15. PRESSURE AND THE FREEZING POINT OF WATER

(*See the drawing.*)

The temperature at which water freezes is the same as that at which it melts. It is simpler to exert pressure on ice than on liquid water, so the effect of pressure on freezing point is best demonstrated by melting ice, rather than freezing water.

You will need: One tall bottle; a length of fine wire (such as fuse wire); two roughly equal weights; an ice cube.

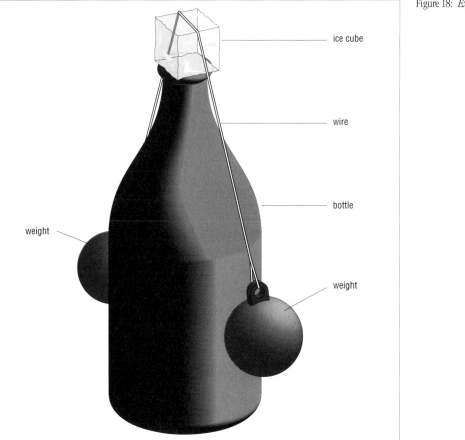

Figure 18: *Experiment 15*

ice cube

wire

bottle

weight

weight

1. Fasten a weight to each end of the wire.

2. Stand the bottle upright and place the ice cube on top of the neck.

3. Carefully place the wire across the cube, with the weights hanging down the side of the bottle.

4. Wait to see what happens. After a short time, the wire will cut through the ice, and the water will refreeze above it. The pressure exerted by the weighted wire is enough to melt the ice.

16. AIR AND DUST PARTICLES

You will need: A vacuum cleaner; a clean piece of white cloth; a magnifying glass (the stronger, the better).

1. Wrap the cloth over the end of the vacuum cleaner hose where the hose attaches to the cleaner. Attach the hose to the cleaner so the cloth is firmly secured.

2. Wait until the weather is dry, and has been dry for a few days. Take the cleaner outdoors or, if the cable will not reach that far, point the end of the hose through an open window.

3. Switch on the cleaner and leave it running for about five minutes.

4. Disconnect the hose and carefully remove the cloth. Lay it on a table in good light and see how much dust it has collected. Use the magnifying glass to try to identify different dust particles. You may find sand grains, pollen, and soil particles.

5. Repeat this procedure, using a fresh cloth, when the weather has been wet for a few days. Compare the difference in the amount of dust you collect.

17. ALBEDO

You will need: Two open-topped boxes, both the same size; two squares of thin cardboard, each large enough to cover the top of a box, one white and one black; a thermometer; enough dry sand to fill both boxes.

1. Fill both boxes with sand and leave them to stand overnight indoors, side by side, so by morning they are at precisely the same temperature.

2. Choose a sunny morning. Take both boxes outdoors and place them on the ground close together, but not touching, where they will be in full sunlight all day.

3. Cover one box with the white card, the other with the black card.

4. About every two hours, measure the temperature of the sand below the surface in each box. Try to measure at the same level in both boxes. Write down the temperatures you measure.

5. Leave the boxes where they are and continue recording the temperature of the sand for an hour or two after sunset.

6. You will find the sand heats faster in the box with the black cover, but cools at the same rate in both boxes.

18. WEIGHING AIR
(*See the drawing.*)
You will need: Two balloons; a straight stick (a bamboo cane is ideal) about 3 feet (90 centimeters) long; string; a pin.

1. Inflate both balloons and tie their ends with string, leaving about 6 inches (15.24 centimeters) of string free.

2. Tie one balloon to each end of the stick.

3. Tie a length of string to the stick, approximately at the stick's center, but make the string loose enough to be moved easily along the stick.

4. Hold the stick and the attached balloons by the central string, and slide this string along the stick until the stick and balloons lie

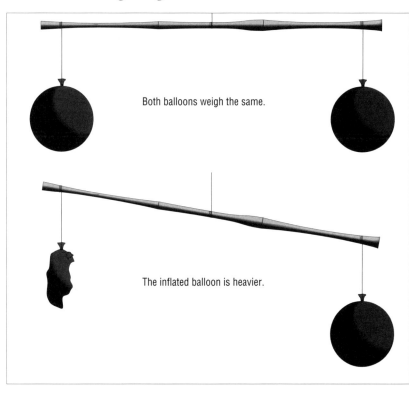

Figure 19: *Experiment 18*

Both balloons weigh the same.

The inflated balloon is heavier.

horizontally. On the scales you have made, the two balloons now balance, which means they weigh the same.

5.   While holding the balanced stick, ask a friend to burst one of the balloons with the pin.

6.   The stick will rise at the end where the balloon burst, because the intact balloon is heavier than the burst one. Since the only difference between them is that one contains air under pressure and the other does not, the inflated balloon must be heavier because of the air it contains.

19. FORMING CLOUD

(*See the drawing.*)

You will need: A 1-gallon (3.8-liter) glass jar; pierced rubber plug; about 6 inches (15 centimeters) of glass tubing; about 18 inches (45 centimeters) of rubber or soft plastic tubing; a strong light.

Figure 20: *Experiment 19*

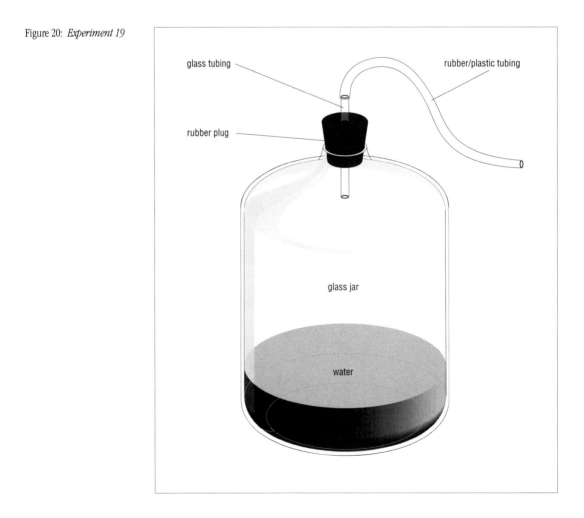

glass tubing

rubber/plastic tubing

rubber plug

glass jar

water

1.   Pour enough water into the jar to fill it to a depth of about 1 inch (2.54 centimeters).

2.   Fit the glass tubing through the plug.

3.   Fit the plug tightly into the neck of the jar.

4.   Fit the rubber or plastic tubing to the outside end of the glass tubing, being sure to make a tight seal between them.

5.   Shake the jar vigorously for a few seconds. This saturates the air inside the jar by encouraging water to evaporate into it, and ensures that the air and water in the jar are at the same temperature.

6.   Blow into the rubber or plastic tube hard for about 15 seconds, then pinch the tube to make sure air cannot escape.

7.   Still pinching the tube, shake the jar again. This saturates the air once more (your breath will have warmed it slightly).

8.   Stand the jar between you and the strong light.

9.   Looking into the jar, release the tube. As the pressure inside falls, the temperature will fall by about 2–3°F (1.1–1.7°C) and water vapor will condense to fill the jar with cloud. The cloud will look very thin, because there is so little of it, but if it filled the room you would think it a dense fog.

20. THE ELECTRIC CHARGE ON WATER MOLECULES

You will need: A glass or plastic rod about 9 inches long (23 centimeters); a piece of nylon cloth.

1.   Turn on a cold water faucet to produce the thinnest steady stream of water you can.

2.   Stroke the rod with the cloth, always in the same direction, to give it an electric charge.

3.   Slowly bring the rod toward the stream of water. As it approaches, notice how the stream bends as the water responds to the charge in the rod.

21. HEAT CAPACITY OF WATER AND SAND

(*See the drawing.*)

You will need: Two containers (they need not be the same size); two thermometers; water; dry sand.

1.   Make sure the sand is really dry. If in doubt, place it at the bottom of the oven while a meal is cooking (it will not harm the food). Then allow it to cool to room temperature.

2.   Fill one container with the dry sand, the other with water.

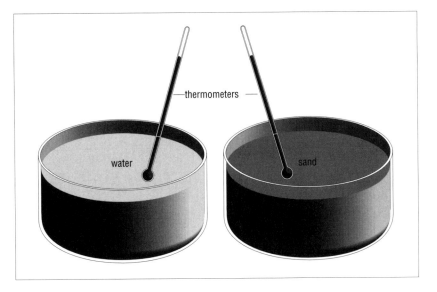
thermometers

water

sand

Figure 21: *Experiment 21*

3. Place both containers in direct sunlight, side by side but not touching.

4. Place one thermometer in each container so the bulb is about 1 inch (2.54 centimeters) below the surface.

5. Check the temperature in each container every few minutes, and note the rate at which they warm.

6. When the temperature ceases to rise any further, move the containers, with their thermometers, to a cool place.

7. Note how quickly each of them cools.

22. MEASURING SNOWFALL

You will need: One tall, straight-sided, plastic bottle; a strip of wood, about 1 × .5 × 40 inches (2.54 × 1.27 × 101.6 centimeters); a ruler; colored masking tape; a craft knife; a pencil that will write on plastic; a ballpoint pen; a large spoon; a spade.

1. Use the craft knife to remove the neck of the plastic bottle, converting it into a straight-sided container.

2. Use the masking tape to mark 1-inch (2.54-centimeter) graduations up the side of the bottle. Label them with numbers, using the pencil.

3. Use the masking tape to mark 1-inch (2.54-centimeter) graduations along the strip of wood. Label them with numbers, using the ballpoint pen.

4. Find a place outdoors as far as possible from buildings, walls, trees, or other obstructions.

5. Push the strip of wood vertically through the snow to the ground, to measure its depth. Note the depth.

6. Use the spoon to scoop snow into the plastic container, being careful not to pack it down. Fill the container up to one of its marks, about three-quarters full. Note the level.

7. If the snow is more than 1 foot (30 centimeters) deep, use the spade to make a vertical cut through the snow. Remove snow from one side of the cut to expose a cross section. Spoon snow from the top, middle, and bottom of the cross section (the snow at the bottom will be slightly compacted by the weight of overlying snow).

8. Take the container with the snow to a warm place indoors, and leave it until all the snow has melted.

9. Note the level of the water, and compare it with the level the snow reached in the container (10 inches [25.4 centimeters] of snow will melt to give about 1 inch [2.54 centimeters] of water).

10. Divide the depth of water by the depth of snow in the container (e.g., $1 \div 10 = 0.1$).

11. Multiply the depth of snow outdoors (measured with the wooden strip) by the fraction you just calculated (e.g., if the depth of snow outdoors is 16 inches [40.64 centimeters], $16 \times 0.1 = 1.6$).

12. The result is the depth of snow converted into its rainfall equivalent.

23. WIND CHILL

You will need: Two watertight containers, one with a lid; a fan; a thermometer.

1. Fill both containers with warm water.

2. Check that the water is at the same temperature in both containers.

3. Place the lid on one of the containers.

4. Set up the fan so it blows unheated air across the surface of the other container.

5. At 10-minute intervals check the temperature of the water in both containers and compare them.

6. Because the "wind" from the fan carries heat away from it, you will find the water in the container exposed to the fan cools faster than the water in the other container although the same air is in contact with both.

24. MEASURING EVAPORATION RATE

You will need: A wide dish; a measuring beaker; a funnel; a clock; a thermometer.

1. Place the dish outdoors in full sunshine.

2. Pour a measured quantity of water into the dish; this should be enough to cover the bottom of the dish to a depth of about 1 inch (2.54 centimeters).

3. Note the amount of water you poured into the dish, and the time.

4. Measure the air temperature close to the dish, but be sure the thermometer is in the shade (it will not give a true reading in full sunlight).

5. After two hours, carefully pour the remaining water from the dish back into the measuring beaker, using the funnel to help. Be careful to spill none.

6. Subtract the amount of water remaining in the dish from the original amount, and divide the result by two. This will tell you the amount of water that has evaporated in one hour.

7. Repeat the experiment, placing the dish in deep shade, in the coolest place you can find. Use your results to relate the rate of evaporation to the temperature.

25. FINDING WATER VAPOR IN DRY AIR AND DRY SOIL

You will need: A glass bottle; an old metal saucepan with a closely fitting lid; a freezer; a kitchen stove.

1. Wait for a really hot, dry day, when the air and ground seem to hold no water.

2. Place the bottle and the saucepan lid in the freezer, and leave them there for one hour.

3. Fill the saucepan with soil to within about 1 inch (2.54 centimeters) of the top.

4. After you have waited an hour, take the bottle from the freezer, place it on a table, and watch it. Within seconds water droplets will form on it. This is moisture taken from the air.

5. Place the saucepan of soil on the stove.

6. Turn the ring or burner to its lowest setting (the aim is to warm the soil, not burn it).

7. Take the lid from the freezer and put it on the pan.

8. Leave the saucepan until the sides feel comfortably warm when you hold them.

9. Remove the lid. Its inside surface will be covered with water droplets. This water has come from the soil you thought was completely dry.

26. SURFACE TENSION OF WATER

You will need: A bowl of clean water; a metal paper clip; a table fork.

1. Leave the bowl of water to stand until the water is quite still and the surface smooth.

2. Lay the dry paper clip across the tines of the fork.

3. Very gently lower the fork into the water. The paper clip will float.

4. Gently push the paper clip below the surface of the water. The paper clip will now sink.

27. PARTICLE SIZE AND POROSITY

(*See the drawing.*)

You will need: A container with a perforated base; two measuring jugs; a watch with a second hand; a quantity of marbles and dry sand; a wax crayon.

1. Pour the marbles into the container, and use the crayon to mark the level they reach on the outside of the container.

2. Place a measured amount of cold water (e.g., a quart [.95 liter]) into one jug.

Figure 22: *Experiment 27*

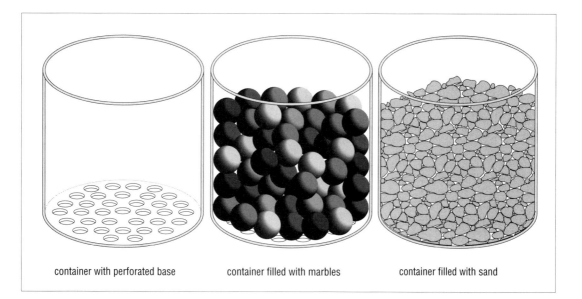

| container with perforated base | container filled with marbles | container filled with sand |

3. Place the container over the second jug, so the jug will catch all the water flowing down.

4. Pour the water from the first jug over the marbles.

5. As soon as water starts to drip into the lower jug, note the time.

6. Measure how long it takes until water stops dripping from the container.

7. Empty out the marbles and dry the container. Empty out all the water.

8. Fill the container with dry sand up to the mark that indicated the level reached by the marbles.

9. Pour the same amount of water over the sand, and time how long it takes before the container stops dripping.

10. You can repeat this experiment using gravel, small stones, and soil.

28. OVERFLOWING (ARTESIAN) WELL

Figure 23: *Experiment 28*   (*See the drawing.*)

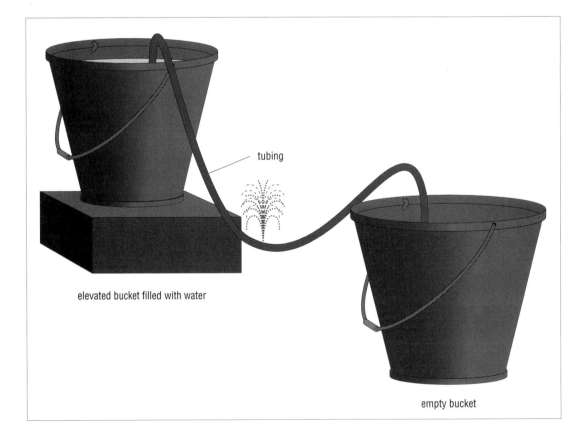

tubing

elevated bucket filled with water

empty bucket

You will need: Two buckets; a length (about 3 feet [90 centimeters]) of rubber or plastic tubing about .5 inch (1.27 centimeters) in diameter; a strong clip; a thumbtack.

1. Fill one of the buckets with water.

2. Place the two buckets so the full one is at a higher level than the empty one (e.g., one on a table, the other on a chair beside the table).

3. Place one end of the tubing into the full bucket, so its end is several inches below the surface, and have someone hold it there, so the end always remains below the surface.

4. Place the other end in your mouth, and suck until water flows from the tubing.

5. Place your tongue across the end of the tubing to halt the flow, then grip the tubing firmly between finger and thumb to seal it.

6. Remove the end of the tubing from your mouth and place it in the lower bucket.

7. Release the end and water will flow (you are siphoning water from the upper to the lower level).

8. Seal the lower end of the tubing with the clip, leaving the end inside the bucket and making sure it stays there.

9. Between the buckets, push the tubing down into a loop approximately level with the bottom of the full bucket.

10. Use the thumbtack to pierce the tubing near the center of the loop (the lowest point). Water will spurt out of the hole, for the same reason water spurts from an overflowing (artesian) well.

29. MEASURING BUOYANCY

(*See the drawing.*)

You will need: One large, watertight box; one large glass dish; one bottle with a cork or screw cap; scales for weighing; water.

1. Weigh the watertight box.

2. Weigh the empty, sealed bottle.

3. Place the glass dish in the box and fill it with water all the way to the brim, being careful not to spill any water over the sides.

4. Gently lower the bottle into the glass dish until it floats. As it goes into the dish, water will spill over the sides into the box.

5. Carefully remove the bottle and set it to one side.

6. Being careful not to spill any more water, remove the dish with its remaining water and set it to one side.

7. Weigh the watertight box with the water it contains.

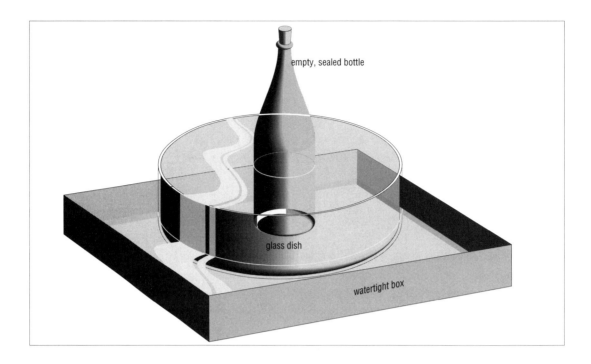

empty, sealed bottle

glass dish

watertight box

Figure 24: *Experiment 29*

8.   Subtract the original weight of the box from the combined weight of the box and water. The difference will be equal to the weight of the bottle.

30. MEASURING RAINFALL

(*See the drawing.*)

You will need: A wide, straight-sided container; a straight-sided clear glass bottle, at least 8 inches (20.32 centimeters) from the base to the shoulder; a funnel; masking tape; a ruler; a ballpoint pen.

1.   Fasten a strip of masking tape to the side of the bottle. Make sure it is straight.

2.   Place the ruler upright in the straight-sided container, and pour in water until it reaches a depth of 1 inch (2.54 centimeters).

3.   Place the funnel in the neck of the bottle.

4.   Pour in the water from the container. Be careful not to spill any.

5.   Note the level the water reaches in the bottle, and mark it with the ballpoint pen on the tape. Label it "1."

6.   Pour another inch (2.54 centimeters) of water into the container, then add it to the water already in the bottle. Mark the new level "2."

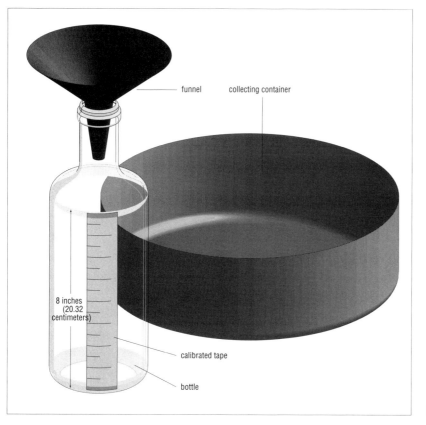

funnel   collecting container

8 inches
(20.32
centimeters)

calibrated tape

bottle

Figure 25: *Experiment 30*

7. Now complete the calibration of the tape, marking in half inches and inches all the way to the top (the shoulder of the bottle).

8. Empty the bottle.

9. To measure the rainfall, place the container outdoors in the open, well away from bushes, trees, and buildings. Always leave it in the same place.

10. Once a day at the same time, empty the container into the bottle. The distance the water reaches will indicate the amount of rainfall since you last checked.

11. In summer or windy weather, you may lose water from the collecting container by evaporation. To prevent this, you can make (or find) a funnel to fit the container. It must be exactly the same size as the top of the container (or it will collect rain from a bigger area and distort the reading). If you cannot find a suitable funnel, you might make one from a piece of plastic, folded and taped.

Wind measurement and cloud classification

Wind Measurement

The first successful system for classifying winds was introduced in 1806 by Admiral Sir Francis Beaufort. Two additional scales were added later to classify winds stronger than those allowed for in the Beaufort scale. Meteorologists of the U.S. Weather Bureau introduced the Saffir/Simpson Hurricane Scale in 1955, and in 1971 T. Theodore Fujita and Allen Pearson introduced the Fujita Tornado Intensity Scale.

THE BEAUFORT SCALE OF WIND STRENGTH

Force 0. 1 MPH (1.6 KPH) or less. Calm. The air feels still, and smoke rises vertically.

Force 1. 1–3 MPH (1.6–4.8 KPH). Light air. Wind vanes and flags do not move, but rising smoke drifts.

Force 2. 4–7 MPH (6.4–11.2 KPH). Light breeze. Drifting smoke indicates the wind direction.

Force 3. 8–12 MPH (12.8–19.3 KPH). Gentle breeze. Leaves rustle, small twigs move, and flags made from lightweight material stir gently.

Force 4. 13–18 MPH (20.9–28.9 KPH). Moderate breeze. Loose leaves and pieces of paper blow about.

Force 5. 19–24 MPH (30.5–38.6 KPH). Fresh breeze. Small trees that are in full leaf wave in the wind.

Force 6. 25–31 MPH (40.2–49.8 KPH). Strong breeze. It becomes difficult to use an open umbrella.

Force 7. 32–38 MPH (51.4–61.1 KPH). Moderate gale. The wind exerts strong pressure on people walking into it.

Force 8. 39–46 MPH (62.7–74 KPH). Fresh gale. Small twigs are torn from trees.

Force 9. 47–54 MPH (75.6–86.8 KPH). Strong gale. Chimneys are blown down; slates and tiles are torn from roofs.

Force 10. 55–63 MPH (88.4–101.3 KPH). Whole gale. Trees are broken or uprooted.

Force 11. 64–75 MPH (102.9–120.6 KPH). Storm. Trees are uprooted and blown some distance. Cars are overturned.

Force 12. More than 75 MPH (120.6 KPH). Hurricane. Devastation is widespread. Buildings are destroyed, and many trees are uprooted. In the original instruction, "no sail can stand."

SAFFIR–SIMPSON HURRICANE SCALE

Category	Pressure at center (inches of mercury)	Wind speed (MPH)	Storm surge (feet)
1	28.94 (73.5 cm)	74–95 (119–153 KPH)	4–5 (1.2–1.5 m)
2	28.5–28.91 (72.39–73.43 cm)	96–110 (154.4–177 KPH)	6–8 (1.8–2.4 m)
3	27.91–28.47 (70.9–72.31 cm)	111–130 (178.5–209 KPH)	9–12 (2.7–3.6 m)
4	27.17–27.88 (69.01–70.82 cm)	131–155 (210.8–249.4 KPH)	13–18 (3.9–5.4 m)
5	below 27.17 (69.01 cm)	more than 155 (249.4 KPH)	more than 18 (5.4 m)

FUJITA TORNADO INTENSITY SCALE

Rating	Wind speed (MPH)	Damage expected
F-0	40–72 (64–115.8 KPH)	Light damage
F-1	73–112 (117.4–180.2 KPH)	Moderate damage
F-2	113–157 (181.8–252.6 KPH)	Considerable damage
F-3	158–206 (254.2–329.6 KPH)	Severe damage
F-4	207–260 (331.2–418.3 KPH)	Devastating damage
F-5	261–318 (419.9–511.6 KPH)	Incredible damage

Cloud Classification

Despite attempts over many centuries to categorize clouds, it was not until 1803 that Luke Howard, an amateur meteorologist, devised a classification that works. It still forms the basis of the modern international system, which allots names to clouds according to their appearance and structure.

Clouds are grouped into 10 distinctive types, called genera (singular *genus*). The genera are subdivided into species (singular and plural), and species into varieties. Genera and species names, which are in Latin, have standard abbreviations; variety names are usually written in full. Stratocumulus (Sc), for example, may form in an almond or lens shape, producing the species lenticularis (len), abbreviated as Sc$_{len}$. If a cloud appears to consist of bands, it may be given the variety name radiatus.

Cloud genera are described as low-level, medium-level, or high-level, according to the height at which they most commonly form, although clouds can form at higher or lower levels. Large storm clouds, which have a low-level base but extend to a great height, are counted as low-level clouds, mainly for convenience. Most medium-level clouds have names beginning with the prefix *alto-*, and the names of high-level clouds have the prefix *cirr-*.

CLOUD GENERA

LOW-LEVEL CLOUDS. Cloud base from sea level to 1.2 miles (1.93 kilometers).

Stratus (St) An extensive sheet of featureless cloud that will produce drizzle or fine snow if it is thick enough.

Stratocumulus (Sc) Similar to St, but broken into separate, fluffy-looking masses. If thick enough, it also produces drizzle or fine snow.

Cumulus (Cu) Separate, white, fluffy clouds, usually with flat bases. There may be many of them, all with bases at about the same height.

Cumulonimbus (Cb) Very large cumulus, often towering to a great height. Because they are so thick, Cb clouds are often dark at the base. If the tops are high enough, they will consist of ice crystals and may be swept into an anvil shape.

MEDIUM-LEVEL CLOUDS. Cloud base from 1.2–2.5 miles (1.93–4 kilometers) in polar regions, 4–5 miles (6.4–8 kilometers) in temperate and tropical regions.

Altocumulus (Ac) Patches or rolls of cloud joined to make a sheet. Ac is sometimes called "wool-pack cloud."

Altostratus (As) Pale, watery, featureless cloud that forms a sheet through which the Sun may be visible as a white smudge.

Nimbostratus (Ns) A large sheet of featureless cloud, often with rain or snow, that is thick enough to completely obscure the Sun, Moon, and stars. It makes days dull and nights very dark.

HIGH-LEVEL CLOUDS. Cloud base from 2–5 miles (3.2–8 kilometers) in polar regions, 3–11 miles (4.8–17.6 kilometers) in temperate and tropical regions. All high-level clouds are made entirely from ice crystals.

Cirrus (Ci) Patches of white, fibrous cloud, sometimes swept into strands with curling tails ("mares' tails").

Cirrocumulus (Cc) Patches of thin cloud, sometimes forming ripples, fibrous in places, and with no shading that would define their shape.

Cirrostratus (Cs) Thin, almost transparent cloud forming an extensive sheet and just thick enough to produce a halo around the Sun or Moon.

Glossary

ablation The disappearance of ice and snow from the surface because it has melted or sublimed (changed directly from solid to gas).

absolute humidity The mass of water present in a unit volume of air, usually expressed as grams of water vapor per cubic meter of air.

absolute vorticity Relative vorticity plus the local effect of the rotation of the Earth about its own axis. *See* vorticity; relative vorticity.

adiabatic Describes a change in the temperature and pressure of a mass of air that occurs without any exchange of energy with the surrounding air. When a "parcel" of air rises, it expands and cools adiabatically; when it descends, it contracts and warms adiabatically.

advection The horizontal transfer of heat that occurs when warm air moves across a cold surface or cold air moves across a warm surface, and heat passes between the air and the surface.

advection fog Fog that forms when warm, moist air moves over a cold land or water surface. The air is cooled by contact with the surface, causing some of its water vapor to condense. *See also* radiation fog.

albedo The reflectivity of a surface to solar radiation. A light-colored surface, such as snow, reflects a high proportion of the light and radiant heat falling on it and has a high albedo; a dark-colored surface absorbs a high proportion of the light and radiant heat falling on it and has a low albedo. Albedo is expressed as the percentage of radiation reflected from the surface, either as a percentage (e.g. 80 percent) or as a decimal (e.g., 0.8).

amphidromic point Where the tide flows into a confined area and circulates around it (counterclockwise in the northern hemisphere and clockwise in the southern hemisphere), the point around which the wave moves. At the amphidromic point there is no rise and fall of water with the tides.

anemometer An instrument for measuring wind speed. Some anemometers measure the pressure against a plate exposed at right angles to the wind, but the most common design consists of small cups mounted at the ends of horizontal arms, which are free to spin about a vertical axis. The pressure on the plate or rotational speed of the spinning cups is converted into wind speed and shown on a dial.

angular momentum The combination of forces acting on any body spinning about its own axis. These include the mass of the body, its rate of spin (measured as angular velocity, or the number of degrees through which the body turns in one second), and the radius of rotation measured from the rotational axis to the farthest point on the body. In the absence of friction and outside forces acting on the body, angular momentum remains constant (is conserved) so that if one of its components change, one or more of the others changes to compensate.

angular velocity The number of degrees through which a body spinning about its own axis turns in one second.

anticyclonic Describes the situation in which a fluid substance (usually air) flows around an area of relatively high pressure, so that pressure increases toward the center. Anticyclonic flow is clockwise in the northern hemisphere and counterclockwise in the southern hemisphere.

aqueduct A channel built to carry water, often over low-lying ground.

aquiclude (aquifuge) A rock that is almost impermeable, so that although the rock may be saturated, water is unable to flow through it.

aquifer Porous rock, gravel, sand, or other granular material below the ground surface that is saturated with water and through which water flows very slowly. *See also* confined aquifer; perched aquifer; unconfined aquifer.

aquifuge *See* aquiclude.

Atlantic conveyor The system of ocean currents that carry cold, dense water along the ocean floor from the edge of the Atlantic sea ice all the way to Antarctica (*see* North Atlantic Deep Water), and replace it with warmer surface water flowing north. This is the most important oceanic mechanism for transferring heat from the equator to high latitudes; it has a strong influence on climates.

Beaufort scale A scale for estimating wind speed by its effect on everyday objects, such as falling leaves, smoke, and trees. The scale was devised in 1806 by Admiral Sir Francis Beaufort, originally to guide sailors as to the amount of sail their ships should carry in various wind conditions. The scale allots a "force" to each wind strength from 0 to 12; five more values were added in 1955 by meteorologists at the U.S. Weather Bureau to describe hurricane-force winds.

Bernoulli effect   The fall in pressure in a fluid passing through a constriction. It was discovered in 1738 by the Swiss mathematician Daniel Bernoulli and explains why aerofoil surfaces generate lift, why hurricane winds can lift roofs from buildings, and why the pressure at the center of a vortex must be lower than the pressure outside.

biological oxygen demand (BOD)   The most widely used measure of water pollution by organic material such as sewage or decaying vegetation. A sample of water is taken and the amount of oxygen dissolved in it measured. The sample is then stored in darkness at constant temperature for several days and the content of dissolved oxygen measured again. The reduction in this amount is due to oxygen consumed by the bacterial oxidation of organic matter and can be used as a pollution measure.

blocking   The effect of a stationary mass of high-pressure air, bringing settled fine weather to the region over which it lies and forcing other weather systems to move around it.

BOD   *See* biological oxygen demand.

Buys Ballot's law   The observation—made in 1857 by the Dutch meteorologist Christoph Buys Ballot—that in the northern hemisphere, winds flow clockwise around areas of high pressure and counterclockwise around areas of low pressure (these directions are reversed in the southern hemisphere). Expressed another way, if you stand with your back to the wind, in the northern hemisphere the area of low pressure is to your left. A few months before Buys Ballot published his observation, the American meteorologist William Ferrel calculated from the laws of physics that this would be the case.

catchment   In Britain, the area from which water drains into a river or ground water system. *See also* divide; watershed.

CCN   *See* cloud condensation nuclei.

cloud condensation nuclei (CCN)   Small particles onto which water vapor will condense readily when the relative humidity approaches 100 percent.

confined aquifer An aquifer that lies beneath a layer of impermeable material. *Compare* unconfined aquifer; *see also* perched aquifer.

contingent drought Drought that results from a prolonged lack of rain in regions where rainfall is highly variable everywhere, but usually high enough to prevent drought.

CorF *See* Coriolis effect.

Coriolis effect (Coriolis force; CorF) The apparent eastward deflection of a fluid that flows away from the equator and westward deflection of one flowing toward the equator. No force is involved; the deflection is wholly due to the rotation of the Earth, the Earth's surface traveling at a different speed from the fluid crossing it. This explains the flow of air around areas of high and low pressure (*see* Buys Ballot's law) and the approximately circular flow of ocean currents. The effect was discovered in 1835 by the French physicist Gaspard de Coriolis.

Coriolis force *See* Coriolis effect.

cyclonic Describes the situation in which a fluid (usually air) flows around an area of relatively low pressure, such that pressure decreases toward the center. Cyclonic flow is counterclockwise in the northern hemisphere and clockwise in the southern hemisphere.

dew point The temperature to which air must be cooled for water vapor to start condensing from it. This temperature varies according to the absolute humidity of the air.

divide The boundary between two drainage basins, such that water drains in one direction to one side of the divide and in a different direction on the other side. In Britain, a divide is sometimes known as a "watershed."

doldrums Areas over the tropical oceans where winds are usually light and variable.

Doppler effect The change in frequency of waves perceived by an observer if the source of those waves is moving toward or away from the observer. As the source approaches, the wave frequency increases; as it recedes, the frequency decreases. The effect was discovered in 1842 by the Austrian physicist Christian Doppler.

drizzle Fairly constant precipitation comprising water droplets smaller than 0.002 inch (0.5 millimeter) diameter.

dust devil A rapidly rotating wind that occurs in dry desert air. Often many dust devils occur close together. They are strong enough to raise dust and sand, extend to heights of more than 5,000 feet (1.525 meters), and may damage buildings.

easterly wave A weak trough of low pressure that occurs in the tropics and shows as a wave on a weather map. As it travels across the ocean it may weaken and disappear, or strengthen, eventually becoming a tropical cyclone.

ecliptic The plane of the Earth's orbit about the Sun.

Ekman effect The result of the effects of surface wind, friction, and the Coriolis effect on the direction of flow of a layer of air or water moving in relation to the layers above and below it. Wind-driven surface currents in the northern hemisphere flow at about 45° to the right of the wind direction. This angle increases with depth until currents below the surface may flow in the opposite direction to those at the surface, the direction forming a spiral with increasing depth. The effect was explained in 1905 by the Swedish oceanographer Vagn Walfrid Ekman.

El Niño *See* ENSO.

ENSO Abbreviation for the combined El Niño–Southern Oscillation, a change in the distribution of atmospheric pressure over the tropics that weakens or reverses the South Pacific trade winds and causes a warm current (El Niño) to flow toward the west coast of South America.

equation of state The equation relating the temperature, pressure, and density of air. Meteorologists use it to calculate one of these when the other two are known. $p = dRT$, where p is pressure, d is density, T is temperature in kelvin (0 K = -273.15°C; 1 K = 1°C), and R is the universal gas constant (8.314 joules per kelvin per mole).

equatorial trough The region of low surface atmospheric pressure around the equator where trade winds from the northern and southern hemispheres meet.

extratropical hurricane A severe storm, with winds of hurricane force, that occurs in a high latitude, far from the tropics. It has many of the characteristics of a tropical cyclone, but forms by a different mechanism.

fetch The distance over which wind blows across the sea, generating waves.

flood peak formula Any one of a number of mathematical formulae that use measurements of rainfall duration and intensity, drainage patterns, and other relevant factors to predict the maximum height flood waters will reach.

front A boundary between two air masses with different characteristics. Fronts move across the surface of land and sea. If the air behind the front is cooler than that ahead of it, the front is said to be a cold front; if the air behind the front is warmer, it is a warm front.

frost point The temperature to which air must be cooled for its water vapor to start forming ice crystals on exposed surfaces. This occurs when the dew point temperature is below the freezing temperature.

F scale *See* Fujita Tornado Intensity Scale.

Fujita Tornado Intensity Scale (F scale) A system that classifies tornadoes on a six-point scale (F-0 to F-5) according to the speed of their winds and the damage they are likely to cause. The scale was devised in 1971 by T. Theodore Fujita and Allen Pearson. (*See* Wind Measurement and Cloud Classification, p. 128.)

Fujiwara effect The orbiting of two tropical cyclones about a common center, which occurs when the cyclones approach within less than 900 miles (1,448 kilometers) of each other. The phenomenon was first observed in 1921 by the Japanese meteorologist Sakuhei Fujiwara.

funnel cloud A funnel-shaped cloud that forms beneath another cloud and extends toward the ground. When it touches the ground, it becomes a tornado.

gas laws The physical laws that describe the relationship between the density, pressure, and temperature of a gas such as air. *See* equation of state.

geostrophic wind A wind that blows almost parallel to isobars; its direction is determined wholly by the pressure-gradient force and Coriolis effect. This wind occurs far enough above the surface to be unaffected by friction with the ground or by obstacles, such as trees or buildings, on the ground.

glacioisostasy The slow rise of the land surface following the melting of a thick ice sheet.

graupel Soft hail, resembling small snow pellets.

greenhouse effect The warming of the lower atmosphere due to the absorption by certain gas molecules of long-wave radiation from the surface. The absorbed heat is then re-radiated, warming other gas molecules.

groundwater Water that has drained downward from the ground surface and saturates a layer of porous rock or particles above a layer of impermeable material.

gyre One of the major systems of currents, found in all oceans, that flow in an approximately circular direction, clockwise in the northern hemisphere and counterclockwise in the southern. Gyres are centered at about 30° north and south of the equator and to the west of the ocean center.

Hadley cell The rise of air in the tropics and sinking of air in the subtropics, forming a convective cell described in 1735 by the English meteorologist George Hadley.

heat capacity The amount of heat required to raise the temperature of a unit mass of a substance by one kelvin ($1°K = 1°C$). Scientists measure the amount of heat in joules and the mass in grams.

humidity mixing ratio *See* mixing ratio.

hygrometer An instrument for measuring atmospheric humidity.

intertropical convergence The region near the equator where the trade winds from the northern and southern hemisphere meet and converge.

invisible drought A period following a drought when the ground is extremely dry and water tables low, so that apparently abundant rainfall fails to recharge aquifers and fill reservoirs, and restrictions on water use introduced during the drought emergency must remain in force.

isobar A line on a weather map joining points where the atmospheric pressure is the same.

isobaric surface A surface where the atmospheric pressure is the same throughout. The height of such a surface will vary, producing shapes resembling hills and valleys.

isohyet A line on a weather map joining points where the amount of rainfall is the same.

isotherm A line on a weather map joining points where the temperature is the same.

jet stream A narrow belt of strong wind blowing in a generally west-to-east direction at speeds of 100–200 MPH (160–320 KPH) or sometimes more at a height of 6–9 miles (9.6–14.4 kilometers). Jet streams form where tropical and polar air meet (the polar front jet stream) and at the high-latitude boundary between sinking tropical and subtropical air (the subtropical jet stream), in each case the jet stream lying inside the warmer air.

kinetic energy The energy a body possesses by virtue of being in motion.

latent heat The heat that must be absorbed to break molecular bonds and convert a substance from a solid to a liquid and a liquid to a gas (e.g., ice to water and water to water vapor). The same amount of heat is released when the molecular bonds reform as a gas condenses to a liquid and a liquid solidifies. Latent heat is absorbed and released without altering the temperature of the substance itself, but is taken from or released into the surrounding medium.

line squall *See* squall line.

Little Ice Age The period from about 1550 to about 1860, during which temperatures all over the world were lower than they had been prior to 1550 or after 1860 and glaciers expanded everywhere. In England, average temperatures in the 1690s were about 2.7°F (1.5°C) lower than in the middle of the 20th century.

mamma One of a number of projections below the base of a cloud, most commonly a big cumulonimbus that may produce tornadoes.

Maunder minimum The period between 1645 and 1715 during which the number of sunspots recorded was much lower than usual; several 10-year periods elapsed when no sunspots at all were seen. This period coincided with the coldest part of the Little Ice Age. Other prolonged episodes of cold climate have since been linked to sunspot minima, and episodes of warm climate to sunspot maxima. The original minimum was described in 1894 by the British astronomer Edward Maunder.

meridional flow Movement of air or water in a northerly or southerly direction (i.e., approximately parallel to lines of longitude, or meridians). *Compare* zonal flow.

mesocyclone A region of rapidly rotating air, up to 6 miles (9.6 kilometers) in diameter, inside a large storm cloud. Rotation starts in the middle of the cloud and extends downward, sometimes all the way to the ground, where it becomes a tornado.

Milankovich theory An explanation for ice ages and the inter-glacials separating them, proposed in 1930 by the Serbian climatologist Milutin Milankovich. He suggested that three cycles affecting the amount of solar radiation the Earth receives at different latitudes periodically coincide. When the effect is to reduce solar radiation to a minimum, the onset of an ice age is triggered, and when the effect is to maximize the solar radiation received, the world climate grows warmer.

millibar One-thousandth of a bar; a widely used unit for reporting atmospheric pressure. 1 bar = 1 atmosphere = 1,000 dynes per square centimeter = 100,000 pascals = 750 millimeters of mercury = 30 inches of mercury.

mixing ratio In a mixture of two gases, the ratio of the mass of one to that of the other. The ratio of the mass of water vapor to the mass of dry air containing it is called the "humidity mixing ratio."

NADW *See* North Atlantic Deep Water.

North Atlantic Deep Water (NADW) Cold, dense water that forms near the edge of the sea ice in the North Atlantic, sinks to the ocean floor, and flows south all the way to Antarctica.

ozone layer A layer of the stratosphere, at a height of 10–20 miles (16–32 kilometers), where ozone (O_3) accumulates, reaching a concentration of up to 10 parts per million.

perched aquifer Flowing groundwater that lies above an aqui-clude.

permanent drought Drought that is due to a generally arid climate, where rainfall is sparse and unreliable.

PGF *See* pressure gradient force.

polar front The boundary in middle latitudes between polar and tropical air. Depressions form along it. The front moves toward the equator in winter and toward the pole in summer.

polar low An area of low atmospheric pressure that forms where very cold air flowing off an ice sheet meets much warmer air over the ocean, producing cumulus clouds and heavy snowstorms.

precipitation Water that falls from the air to the ground, regardless of its form. The term includes drizzle, rain, snow, graupel, hail, and also mist and fog.

pressure gradient force (PGF) The force acting on air in the direction of a center of low atmospheric pressure, i.e., at right angles to the isobars surrounding the center. Its strength is proportional to the rate at which the pressure changes with horizontal distance; i.e., the pressure gradient.

psychrometer An instrument for measuring humidity using a wet-bulb and dry-bulb thermometer.

radiation fog Fog that forms in still air when the ground surface has cooled overnight by radiating away the heat it absorbed by day. Contact with the chilled ground cools the surface layer of air to below its dew point temperature, causing water vapor to condense. *See also* advection fog.

radiosonde A package of instruments for taking measurements of atmospheric temperature, pressure, and humidity that is carried aloft by a balloon and transmits its readings by radio.

rawinsonde A radiosonde that also carries a radar reflector allowing it to be tracked from the ground to provide data on wind speed and direction at the heights it reaches.

relative humidity The amount of water vapor present in the air expressed as the percentage of the amount that would saturate the air at that temperature and pressure.

relative vorticity The amount of rotation of a fluid about an axis (usually a vertical axis) in relation to the surface of the Earth. *See* absolute vorticity.

ridge An area of high atmospheric pressure with an elongated shape; on an isobaric map it would stand above the surface like a ridge on a landscape.

riprap Large boulders or concrete blocks that are laid against a sea wall, dam, or other surface exposed to the sea in order to absorb the energy of waves and so protect the structure from erosion.

Rossby waves Undulations, up to 1,500 miles (2,413 kilometers) from crest to crest, that develop in the jet streams. Similar waves occur in the oceans. They were discovered and explained in 1940 by the Swedish–American meteorologist Carl-Gustav Rossby.

Saffir–Simpson scale A five-point scale for reporting the severity of tropical cyclones (hurricanes) devised by meteorologists at the National Oceanic and Atmospheric Administration. (*See* Wind Measurement and Cloud Classification, p. 128.)

scattering The deflection of electromagnetic waves (e.g., sunlight) by gas molecules, particles, and irregular surfaces.

seasonal drought Drought characteristic of regions with dry and rainy seasons that occurs when the dry season is drier or longer than usual.

sleet Ice pellets, no bigger than 0.04 inch (.1 centimeter) in diameter, that do not melt before reaching the ground. In Britain, sleet is a mixture of snow and rain.

snow blitz The rapid onset of an ice age that might occur if, in some areas, the winter snow failed to melt during the summer. The surviving snow would reflect solar heat and light, keeping the ground cool, and the following winter more snow would accumulate and the area of year-round snow would increase.

specific humidity The ratio of the mass of water vapor to the mass of air including water vapor containing it.

squall line (line squall)   A series of storms arranged in a line ahead of a cold front.

stratopause The upper boundary of the stratosphere at a height of about 30 miles (48 kilometers), separating the stratosphere from the mesosphere above. At the stratopause, the average temperature is about 32°F (0°C); above it, temperature decreases with height.

stratosphere The layer of the atmosphere at heights between about 6 and 30 miles (9.6–48 kilometers), which is bounded by the tropopause below and stratopause above and contains the ozone layer. In the lower stratosphere, temperature remains constant with height, at about -76°F (-60°C), but in the upper stratosphere, the absorption of ultraviolet radiation by oxygen and ozone increases the temperature to about 32°F (0°C) at the stratopause.

sublimation The direct change of a substance between solid and gas, without passing through a liquid phase.

suction vortex A miniature tornado that forms in the turbulent air around the edge of a large tornado. Suction vortices follow circular paths around the main tornado, usually for less than one complete revolution before dying, but generate winds up to 100 MPH (160 KPH) faster than those of the main tornado.

supercell A violent updraft of air formed when several convective cells merge inside a large cumulonimbus storm cloud. Within a supercell, air may be rising at up to 100 MPH (160 KPH), and the cell may break through the tropopause to a height of more than 10 miles (16 kilometers). Supercells last much longer than ordinary convective storm cells, and they can trigger tornadoes.

temperature inversion A layer of the lower atmosphere in which temperature remains constant or increases with height. Air rising from below by convection cannot penetrate the inversion, because it comprises air at the same or lower density than the rising air, so air becomes trapped beneath the inversion, together with any smoke or other pollutants it carries.

thermal wind The change in speed and direction with height of the geostrophic wind within a given layer of the atmosphere, due to the change in temperature. In the northern hemisphere the wind blows with the cold air to its left (and to the right, in the southern hemisphere) at a speed proportional to the rate of temperature change (the temperature gradient).

trade winds The prevailing winds that blow to either side of the equator, from the northeast in the northern hemisphere and from the southeast in the southern hemisphere.

tropopause The boundary between the lower (troposphere) and upper (stratosphere) layers of the atmosphere, marking the level at which temperature ceases to decrease with height. Its altitude varies according to the temperature of the sea surface and with the seasons, but averages 6–7 miles (9.6–11.2 kilometers) over the poles (sometimes lower) and about 10 miles (16 kilometers) over the equator.

troposphere The lowest layer of the atmosphere, extending from the surface to the tropopause, within which temperature decreases with height.

trough An area of low atmospheric pressure with an elongated shape; on an isobaric map it appears as a depression in the surface.

unconfined aquifer An aquifer that is not bounded on its upper side by a layer of impermeable material. Water can drain freely into it from above.

vorticity A measure of the rotation of a fluid; in meteorology, the term usually refers to relative vorticity about a vertical axis. *See also* absolute vorticity; relative vorticity.

Walker circulation The west-to-east airflow in the upper troposphere over the tropical Pacific Ocean that acts as a counterflow to the low-level east-to-west airflow caused by the Hadley cell circulation and trade winds. It was discovered in 1923 by Sir Gilbert Walker.

wall cloud A region of rotating cloud below the main base of a large storm cloud. It often releases little or no rain, but tornadoes may be forming in the cloud behind it.

watershed The area within which surface water drains into a particular groundwater and river system. A divide separates one watershed from another. In Britain, a watershed is often called a "catchment," and a divide is often called a "watershed."

waterspout An intense vortex of air that forms over water. It closely resembles a tornado and in some cases is a tornado that formed over land and subsequently crossed water. Waterspouts that form over water do not require a mesocyclone for their development and are smaller and weaker than true tornadoes. Most occur over shallow water in the tropics.

water table The upper boundary of groundwater, below which the soil is fully saturated.

wet bulb–dry bulb thermometer An instrument used to measure the evaporation of water, from which the relative humidity and dew point temperature of the air can be calculated or, more usually, read from tables. The instrument comprises two thermometers: The dry bulb thermometer reads the air temperature. The bulb of the wet bulb thermometer is wrapped in muslin, part of which is immersed in distilled water. The temperature indicated on the thermometer is depressed, because latent heat to evaporate water from the muslin is drawn from the bulb itself. The difference between the dry bulb and wet bulb temperatures is called the "wet bulb depression" and varies with the rate of evaporation, which in turn is determined by the humidity of the air. Relative humidity and dew point temperature are calculated from the wet bulb depression.

zonal flow A movement of air or water in a generally east-to-west or west-to-east direction, approximately parallel to lines of latitude. *Compare* meridional flow.

Further reading

SITES ON THE WORLD WIDE WEB

Colorado State University—Tropical Meteorology
Provides general information on tropical weather, including severe weather.
http://typhoon.atmos.colostate.edu/

Explores! Tropical Hurricane Information
Lists the names of tropical cyclones and supplies details about recent ones.
http://www.met.fsu.edu:80/explores/tropical.html

The Florida Everglades
One of the world's most important wetland areas.
http://www.goflorida.com/south/everglades/info/

Hurricane Tropical Data
Supplies tracks of past hurricanes.
http://thunder.atms.purdue.edu/hurricane.html

SIMS Hurricane Watch
Provides links to other sites. Although the home page says it is updated every few hours, it is not.
http://sc.net/links/hurricanes/

Tropical Cyclone Information
Lists the most intense, deadly, and costliest hurricanes, and also provides tracking charts for them.
http://nhc-hp6.nhc.noaa.gov/hurricane_info.html

Tsunamis
Includes details of past tsunamis, what they are and how they happen, and warning networks.
http://www.geophys.washington.edu/tsunami/welcom e.html

University of Hawaii—Severe Weather
Regularly updated information on the weather of the Pacific.
http://lumahai.soest.hawaii.edu/Severe_Weather/severe.htm l

The Why Files
An educational resource, with links to useful and interesting articles about tornadoes, among many other topics. It loads quickly.
http://whyfiles.news.wisc.edu

BOOKS

Allaby, Michael. *Elements: Air.* New York: Facts On File, 1992.

Allaby, Michael. *Elements: Water.* New York: Facts On File, 1992. Two books in the *Elements* series that provide wide-ranging accounts of every aspect of air and water, including climate, written in nontechnical language for older readers.

Allaby, Michael. *How the Weather Works.* New York: Readers Digest, 1995. A book for younger readers, with many simple experiments you can perform at home.

Barry, Roger G., and Richard J. Chorley. *Atmosphere, Weather and Climate.* New York: Routledge, 1993. A more technical but classic textbook, now in its sixth edition, suitable for older students who wish to study climate in more detail.

Clark, Champ. *Planet Earth: Flood.* New York: Time-Life Books, 1983.

Hidore, John J., and John E. Oliver. *Climatology: An Atmospheric Science.* New York: Macmillan Publishing Company, 1993. A textbook on weather, but written in straightforward language with clear and simple diagrams.

Lamb, H. H. *Climate, History and the Modern World.* New York: Routledge, 2d edition 1995. A full account of the history of climate, its effects on human societies, and ways it may change in years to come, written in a very readable form by a highly distinguished climatologist.

McIlveen, Robin. *Fundamentals of Weather and Climate.* New York: Chapman and Hall, 1992. A textbook for more advanced students, in this case supplying much of the mathematics meteorologists use.

Page, Jake. *Planet Earth: Arid Lands.* New York: Time-Life Books, 1984. These books in the *Planet Earth* series provide simple explanations and many stories about actual instances of the phenomena they describe, with dramatic illustrations.

Simons, Paul. *Weird Weather.* New York: Little, Brown and Company, 1996. A collection of highly entertaining stories about extraordinary weather events.

Whipple, A. B. C. *Planet Earth: Storm.* New York: Time-Life Books, 1982.

Wood, Robert W. *Science for Kids: 39 Easy Meteorology Experiments.* Blue Ridge Summit, Penn.: Tab Books (McGraw-Hill), 1991. As the title indicates, a set of 39 simple experiments you can perform, all illustrated and clearly described.

ARTICLES

Bluestein, Howard B. "Riders on the Storm" in *The Sciences* (March/April 1995): 26–30. New York Academy of Sciences, New York. A firsthand account of stalking tornadoes, written by someone who does it for a living.

Fryer, Gerard. "The Most Dangerous Wave" in *The Sciences* (July/August 1995): 38–43. New York Academy of Sciences, New York. A nontechnical explanation of tsunamis.

Weiss, Harvey. "Desert Storm" in *The Sciences* (May/June 1996): 30–36. New York Academy of Sciences, New York. An account of the drought that triggered the collapse of an ancient civilization.

Index

Italic numbers indicate illustrations.